OUTDOOR ENCYCLOPEDIAS

THE WILDLIFE WATCHING ENCYCLOPEDIA

BY LAURA PERDEW

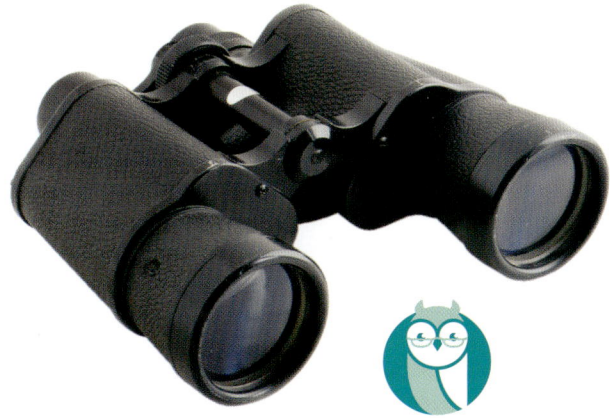

Encyclopedias

An Imprint of Abdo Reference
abdobooks.com

TABLE OF CONTENTS

WHAT IS WILDLIFE WATCHING? 4
Benefits of Wildlife Watching 8

TYPES OF WILDLIFE WATCHING............. 14

WHAT TO BRING.. 24
Food and Water..25
Clothing..30
Navigation 35
Field Guides...39
Binoculars 41
Other Essentials 43

WILDLIFE WATCHING ETIQUETTE 48
Don't Feed the Animals...........................56
Pollution 59
Invasive Species and Disease................63
Protecting Habitats 67
Respecting Other Wildlife Watchers...79

TIME AND LOCATION................................ 80
Location...81
Time of Day 86
Weather 91
Mating and Breeding Seasons........... 104
Migration Patterns 109

FINDING AND IDENTIFYING ANIMALS 112
- Tracks.. 115
- Scat... 124
- Other Signs................................ 129
- Calls and Sounds 136
- Identifying an Animal 143
- Learning More........................... 147

WILDLIFE WATCHING SAFETY............... 150
- Getting Lost............................... 153
- Weather Conditions 155
- Elevation.................................... 162
- Cold.. 163
- Heat.. 166
- Understanding Animal Behavior...... 167
- What to Do If Approached, Attacked, or Bitten............................... 170

GLOSSARY 188

TO LEARN MORE......................... 189

INDEX 190

PHOTO CREDITS 191

WHAT IS WILDLIFE WATCHING?

Animals that are not pets are considered wildlife. This includes mammals, birds, reptiles, amphibians, fish, insects, spiders, and more. Wildlife can be huge, such as grizzly bears or moose. Or it can be small, such as geckos or ladybugs.

Wildlife can be found in every ecosystem. Deserts, tundras, and rain forests have wildlife. Wildlife is also found in water. There is wildlife in rivers, streams, ponds, lakes, wetlands, and oceans. Wildlife lives in parks, backyards, and cities too.

Wildlife watching is a great way to spend time outdoors.

Whale watching tours offer opportunities to see these large animals in their natural habitats.

Wildlife watching is a great way to learn about animals and ecosystems. The type of wildlife to watch depends on the ecosystem and how close it is to a developed area. Urban wildlife might include squirrels, pigeons, and other small animals adapted to city life. Raccoons, rabbits, and birds are often seen in backyards and neighborhoods. Local parks

DID YOU KNOW?

Insects are a type of wildlife that is easy to find. Some people plant gardens to attract butterflies and bees, which drink nectar from flowers. These insects are attracted to gardens that include a variety of native flowers and do not use chemicals that are harmful to insect life.

WHAT IS WILDLIFE WATCHING?

Setting up a bird feeder to attract wildlife is one way to watch animals close to home.

and refuges often have more wildlife to watch. Some may have deer, herons, or prairie dogs. People who want to go deeper into the wilderness have the chance to view even more wild animals. In forests, there may be wild turkeys, foxes, and porcupines. On the ocean, people can go whale watching. In the mountains, wildlife watchers may see bighorn sheep and small rodent-like mammals called pikas. Deserts may offer glimpses of jackrabbits and lizards.

Wildlife watching is an activity anyone can enjoy. It is for the young and the old. It is for people in cities and in rural areas. People can often watch wildlife from their homes. National, state, and local parks offer wildlife watching opportunities that are suited for a wide range of ages and abilities. Many have trails and tours that are accessible to people with disabilities.

BIRD FEEDERS

Bird feeders are an excellent way to watch wildlife at home. There are many kinds of feeders and bird food. Each attracts different birds. For example, cardinals prefer tray or platform feeders they can stand on to eat. Woodpeckers like suet feeders. These feeders hold blocks of animal fat mixed with nuts, fruits, or insects. Hummingbirds drink from brightly colored feeders filled with sugar water. Feeders must be kept clean to prevent birds from getting sick.

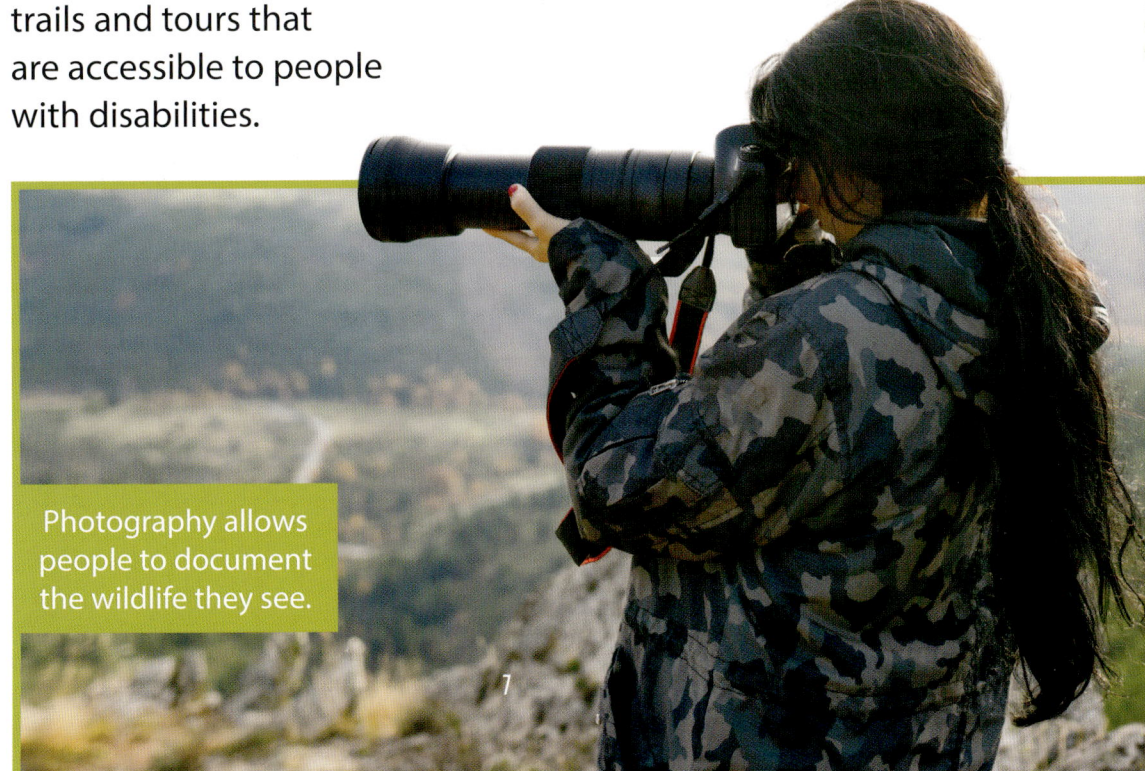

Photography allows people to document the wildlife they see.

WHAT IS WILDLIFE WATCHING?

BENEFITS OF WILDLIFE WATCHING

Watching wildlife benefits physical health. It may involve exercise, like walking or paddling. Carrying gear adds weight to a backpack. This works and tones muscles. Physical activity increases overall strength. It improves heart health and boosts the immune system. It lowers the risk of other medical problems. Wildlife watching can also reduce fatigue and improve sleep.

Kayaking can improve strength in the arms and back.

Studies show that seeing nature and listening to sounds of nature improve mental health.

Wildlife watching is good for mental health too. Research shows that people feel more relaxed after spending time in nature. It reduces symptoms of stress, anxiety, and depression. It improves mood. Time in nature is also linked to improved confidence and self-esteem.

Other studies show that spending time in nature improves brain function, including concentration. People manage their emotions better. They also report feeling more energized, calm, and happy. This overall improved sense of well-being continues even when people return indoors.

WHAT IS WILDLIFE WATCHING?

Wildlife watching is also an opportunity to learn about nature firsthand. It allows people to connect with nature by learning to observe and understand animal behavior. This can reduce fears about wild animals. And it promotes empathy toward animals. This understanding and awareness helps people care for and appreciate nature. They are more likely to take care of

Wildlife watchers learn to understand animal behavior. For example, a squirrel may freeze in place and then scamper up the nearest tree when it feels frightened.

Conservation efforts have helped restore populations of gray wolves in Wyoming's Yellowstone National Park.

DID YOU KNOW?

Studies have shown that even watching wildlife videos is good for one's well-being. There are an increasing number of live wildlife cams. People can watch grizzlies catching salmon in Alaska, puffins nesting in Maine, and manatees underwater in Florida.

natural spaces and support conservation efforts.

Wildlife watching allows people to contribute to citizen science. Citizen science involves people who are not scientists collecting information to be used for scientific research. Volunteers observe the natural world around them. They collect data or take photos. The data and photos are then used and analyzed by scientists.

WHAT IS WILDLIFE WATCHING?

The Fernald Preserve in Ohio hosted a monarch butterfly tagging event in September 2022. This citizen science project helped scientists gather data about how climate change affects monarch migration.

The Monarch Watch Tagging Program is a citizen science project. Participants help scientists tag and record information about monarch butterflies. Each tag tracks the location of an individual butterfly. When citizen scientists put a tag on the butterfly, they also record information about the monarch, such as its size and the date it was tagged. Monarch butterflies migrate. The tags track information about how far a monarch butterfly travels in its life span. This helps scientists learn about how different characteristics of monarchs affect migration.

The Whale Alert app is another example of citizen science. This app is used on the ocean. When people see or hear a whale, they can record the location on the app. Whale Alert was created to protect whales from being hit by boats.

Wildlife watching is an activity that anyone can enjoy. Animals can be seen in yards, parks, and remote wilderness areas. A successful wildlife watching trip requires preparation and planning. People should choose an option that matches their physical ability and experience. In addition, an enjoyable trip requires the correct gear. Wildlife watchers must know how to respect the animals around them. They also need to be prepared with information about how to locate and identify animals and how to stay safe outdoors.

The environmental group National Audubon Society has hosted a Christmas bird count every year since 1900. Volunteers share information about the birds they see, which helps scientists study bird health.

TYPES OF WILDLIFE WATCHING

There are many types of wildlife watching trips. Some are close to home. Other people choose to travel farther away. Trips may last for short periods of time. Or a trip may last days or weeks. Planning a wildlife watching trip includes considering the physical ability and experience level of everyone in the group.

Wildlife can be found in neighborhoods and residential areas.

Swamp tours in Louisiana allow wildlife watchers to view alligators up close.

Some wildlife watching trips may require challenging hikes to remote areas. But other trips are not physically demanding. Wildlife watching can be done off the side of a road, a few steps down a trail, on a beach, or on a dock. Boat and driving tours offer other opportunities for wildlife watching without much physical effort.

Some wildlife watching trips are better suited for beginners. Guides can help point out and identify animals. Other trips may require more experience. People on these trips may be expected to have a solid background in locating wildlife themselves.

TYPES OF WILDLIFE WATCHING

Wildlife watchers who take self-guided trips plan and prepare for the trip themselves. They choose their own locations and decide when to go and how long the trip will last. Self-guided tours allow wildlife watchers to move at their own pace. They offer more flexibility than guided tours. But this independence means that people are responsible for packing the proper gear for their trips. They also need to know how to handle emergencies on their own.

Wildlife watching from a boardwalk does not require much physical intensity.

A guide spoke to passengers about the marine life they might see off the coast of California in 2018.

Guided tours are another option for wildlife watching. These tours are led by experts who are knowledgeable about the wildlife and the location. When selecting an option for a guided tour, people should pay attention to the amount of physical activity the tour requires. They should also make sure the tour is a good fit for their experience level. The guide or company does the planning. Guides are prepared to handle emergencies. Most guided tours are done with a group of people. The pace of travel is also set by the group or guide. There is little flexibility in the schedule compared with self-guided tours.

TYPES OF WILDLIFE WATCHING

Some guided tours in national and state parks are available to the public for free. Tours in these locations are often led by a ranger who knows the region well. Other guided tours cost money. They can be great options for wildlife watchers who want to travel to locations that are dangerous, remote, or otherwise difficult to access.

For example, in Maui, Hawaii, whale watching tours are available between November and May. Humpback whales migrate to Hawaii's warmer water for the winter. Guides take groups out on the ocean and provide a chance to see whales up close. They share information about the animals.

Park rangers can answer questions about trails and wildlife viewing.

National parks are popular destinations for wildlife watching. Wildlife watchers should find out whether there are entrance fees.

In the Everglades in Florida, wildlife watchers ride on a tram to explore the wetland habitat. Guides provide information about the habitat. They point out and teach those on tours about wildlife in the area, including alligators.

It is important for wildlife watchers to check park and wilderness area guidelines before a visit. These places may require permits or entry passes. Visitors should also look for information about the best time

TYPES OF WILDLIFE WATCHING

to visit. Parks and trails may be crowded during certain times of the year. It may be difficult to travel to remote parks when it is not peak visiting season because roads may not be as well maintained. These factors can be challenges to watching wildlife. On the other hand, national parks and wilderness areas offer opportunities to see a great variety of animals and habitats. National parks attract millions of visitors each year. People who travel to remote locations

People should drive slowly when there are bison on the road or wait for the animals to move out of the way. They should not honk their horns.

within these protected areas find quiet, solitude, and wildlife that is not disturbed by human development.

Many parks and refuges have wildlife drives that allow people to see animals from their vehicles. Seeing wildlife on these drives is not guaranteed, but the routes are along areas that animals are known to visit. The Wildlife Loop Road in South Dakota's Custer State Park offers visitors the chance to see herds of bison from the safety of their cars. Bighorn sheep,

TYPES OF WILDLIFE WATCHING

Wildlife watchers may spot bighorn sheep from their cars when driving on Badlands Loop Road in Badlands National Park in South Dakota.

mountain goats, and grizzly bears are some of the wildlife that can be seen on the Going-to-the-Sun Road in Montana's Glacier National Park.

Wildlife watching does not occur only on land. Freshwater and marine habitats are also full of wildlife. People can take trips to lakes, rivers, or oceans to watch animals. After arriving

DID YOU KNOW?

Snorkeling is a unique way to watch wildlife. With the right gear, snorkelers can see a wide range of marine wildlife. Tropical fish, sea turtles, and coral are some of the wildlife beneath the water's surface.

at their destinations, people can travel by kayak, raft, canoe, or a larger boat. These trips allow individuals to see aquatic wildlife as well as birds and other animals that visit the shorelines. Visitors have opportunities to see wildlife that they may not be able to find on land.

Snorkeling allows people to see underwater wildlife.

WHAT TO BRING

Wildlife watching trips require packing the right gear. The type and amount of gear varies depending on the length, location, and type of trip. Food, water, and clothing are essentials for any outing. Trips into wilderness areas require maps and navigation tools. These items and many others increase the success and safety of the excursion.

Before heading out on a wildlife watching trip, people should make sure they have all the gear they need.

Trail mix has many nutrients and can provide a quick boost of energy.

FOOD AND WATER

People should pack snacks and meals to last the duration of the outing. They should bring extra food in their packs, especially for trips into the backcountry. People can use up a lot of calories while wildlife watching. They will need to pack more food for trips that include challenging hikes or other intense activities. Extra food should also be packed in case of emergencies. Storms and injuries may prolong a trip. The extra food will provide energy if a trip takes longer than expected.

Foods need to be able to withstand time in a backpack. The best types of foods to pack are rich in proteins, fats, and complex carbohydrates. These foods provide the body with

WHAT TO BRING

energy that lasts for a long period of time. Trail foods such as energy bars, trail mix, nuts, and jerky have these nutrients. Cheeses, meats, and sandwiches also provide a lot of energy. Consuming food regularly throughout a trip keeps the body fueled.

Foods with a lot of sugar should be avoided. These foods give an initial boost of

DID YOU KNOW?

Weight, activity level, and other factors affect calorie requirements. On average, people burn between 200 and 600 calories an hour when being active outdoors.

When on a wildlife watching trip, make sure to pack out and properly dispose of food packaging or other food waste.

energy, but the energy is used up quickly. This leads to fatigue or sluggishness. People should avoid sodas, candy, sugary breakfast cereals, prepackaged cookies, and fruit juice concentrates.

WHAT TO BRING

Staying hydrated is also important. The amount of water a person needs depends on the intensity of the trip. For more challenging trips, including those in hot weather or at high altitudes, more water is necessary. On average, people should plan to drink about 2 cups (0.5 L) of water every hour. People need to remember to drink water regularly, even on

People should drink water regularly during a wildlife watching trip, even if they do not feel thirsty.

Drinking directly from natural water sources can cause serious illness.

cold days. This is more effective at preventing dehydration than consuming a lot of water at once. Electrolyte tablets and powders are also available to mix with water. Electrolytes replace nutrients the body loses through sweat.

    A water purifier or filter is essential for long trips and travel to remote areas. If an outing is extended due to an emergency, packed water may run out. Even if a lake or stream looks clean, the water may be contaminated with viruses and bacteria. Drinking untreated water can make people sick. Water treatment gear allows individuals to safely drink water from natural sources. People should practice using a purifier or filter at home before taking the gear on a trip.

WHAT TO BRING

CLOTHING

Dressing in layers is important when spending time outdoors. Layers allow people to stay comfortable as temperatures change throughout the day. They can be taken off when temperatures rise or during high levels of activity. Or layers can be put on if the temperature falls or when people cool down while resting. Clothing should be lightweight and quick drying. Clothes can become damp from sweat or rainfall. Wet clothing pulls heat away from the skin, which can cause chills or hypothermia. Nylon, polyester, and wool fabrics are quick drying. Denim and cotton should be avoided as these materials absorb moisture and take a long time to dry.

Layers of clothing can be put on or taken off as weather conditions change throughout the day.

Hiking boots provide support for the ankles and prevent injuries to the feet.

People should dress for comfort when wildlife watching. Loose and stretchy clothing allows individuals to move freely. Comfortable footwear is crucial. Hiking shoes and boots should be broken in prior to the trip to prevent blisters. Shoes need to be appropriate for the terrain. Trips in steep, rocky areas may require footwear with ankle support and slip-resistant treads.

WATER GEAR

In addition to the gear required for a land-based trip, people who are wildlife watching on the water need additional gear. They should pack extra clothing to change into in case they get wet. A waterproof pack protects gear from getting damp. While on the water, everyone should wear a life vest or jacket.

WHAT TO BRING

A raincoat and other waterproof clothing help keep people warm and dry.

A waterproof raincoat is essential for all weather. Raincoats offer protection against rain and snow. They provide a buffer against wind. They hold in body heat. In addition to being waterproof, a raincoat should also be breathable. This means the raincoat allows moisture from sweat to escape so it does not build up inside the coat.

Additional gear may be necessary when wildlife watching in certain weather conditions. In cold weather, people should pack a warm jacket. A knit hat, gloves, and a scarf also help keep a person warm. Other gear should be worn during sunny trips. A wide-brimmed hat protects the face, neck, and eyes from the sun.

Clothing color is another factor to consider for wildlife watchers. In most cases, clothing that blends in with the environment is best. Colors such as browns, grays, and greens blend in with most habitats. By wearing these colors, people are less visible to animals. They have less of an impact on animals' activities, which increases the chances of seeing wildlife. However, other animals may be attracted to brightly colored clothing. For example, people wanting to view hummingbirds should wear reds, yellows, and oranges.

Wearing warm gear in cold weather protects people from harmful health conditions such as hypothermia.

WHAT TO BRING

People should also be aware of how much noise their clothing makes. Fabrics may make sound when they rub against each other. Shoes may squeak. Buckles and zipper tabs may jingle. Sudden or loud noises may scare off wildlife.

Some wildlife photographers and wildlife watchers wear camouflage to avoid disrupting natural animal behaviors.

It can be helpful to have multiple navigation options on hand.

NAVIGATION

People who plan to watch wildlife in parks or wilderness areas should always have a map. Maps are often available online. They can be printed or saved to a phone before leaving home. Most local, state, and national parks have maps available at visitor centers. Maps are often posted at trailheads too. People can take a photo to use along the way.

WHAT TO BRING

It takes practice to learn how to read and use a topographic map.

Outdoor recreation stores often have topographic maps available for purchase. These maps provide details about the topography, or landscape, of an area. For example, they show locations of streams, rivers, and lakes. They also indicate the steepness of the terrain. These maps include information about the vegetation, such as which areas are forested and which are more open. They also show human-made features, such as trails, picnic areas, and more. This information helps wildlife watchers avoid getting lost.

Many wildlife watchers use a compass and a map for navigation. Traveling to remote areas requires a basic understanding of how to use a compass. A compass has a needle that points north. People can look at the direction the needle is pointing to know which way they are facing. They can use this information to help orient themselves when looking at a map.

A compass should be held flat in order to get an accurate reading.

WHAT TO BRING

Other wilderness travelers use a Global Positioning System (GPS). A smartphone has a GPS that determines a person's location. It can also give directions to a destination. People can download maps to their phones before a trip if they know they will be traveling to an area without cellular service. They can use the downloaded maps and the phone's GPS to navigate. Another option is a GPS device that uses satellite signals. These devices remain effective in remote locations. Many GPS devices have a digital screen that displays a map. An individual who plans to use a GPS in the backcountry should practice close to home first. In addition, users should carry spare batteries. A compass and map are still recommended as backups in case a smartphone or GPS device malfunctions, loses signal, or runs out of power.

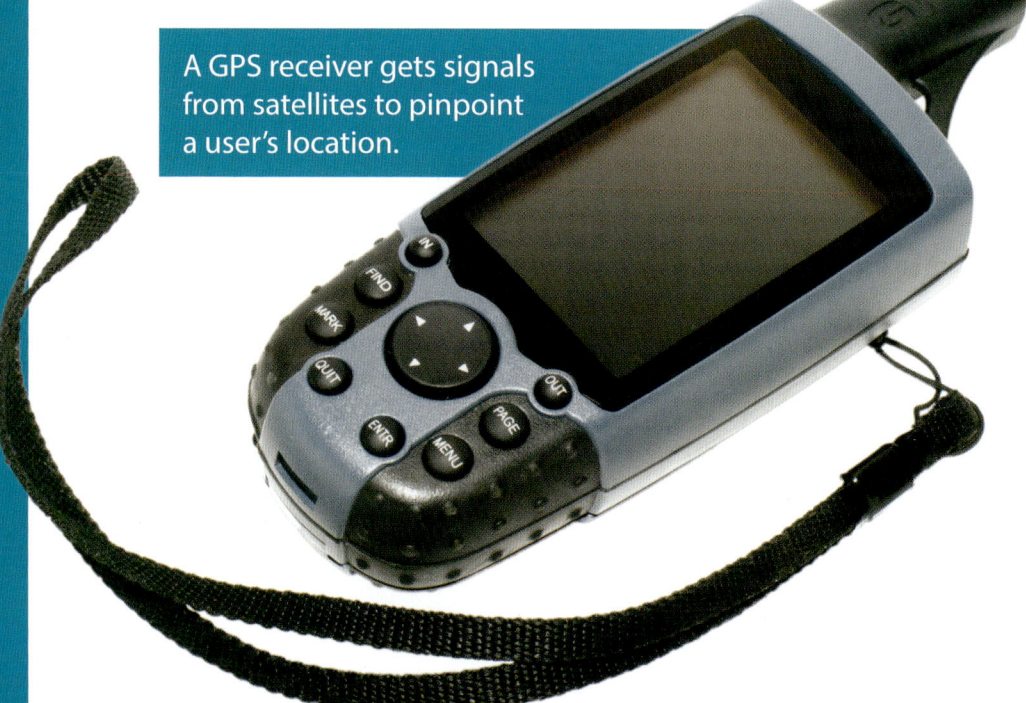

A GPS receiver gets signals from satellites to pinpoint a user's location.

Field guides often have pictures that help with wildlife identification.

FIELD GUIDES

Field guides assist with locating and identifying wildlife. These guides are available in a variety of forms. A field guide can be a book, a brochure, or an app. Some provide information about specific wildlife groups, such as the birds, reptiles, or amphibians in a region. Others contain more general information about common animals in an area. People can also learn about wildlife from quick reference guides. These guides are only one or two pages long, making them easy to carry on a wildlife watching trip. Most national, state, and local parks have free brochures for visitors. These brochures often include information about wildlife in the area.

WHAT TO BRING

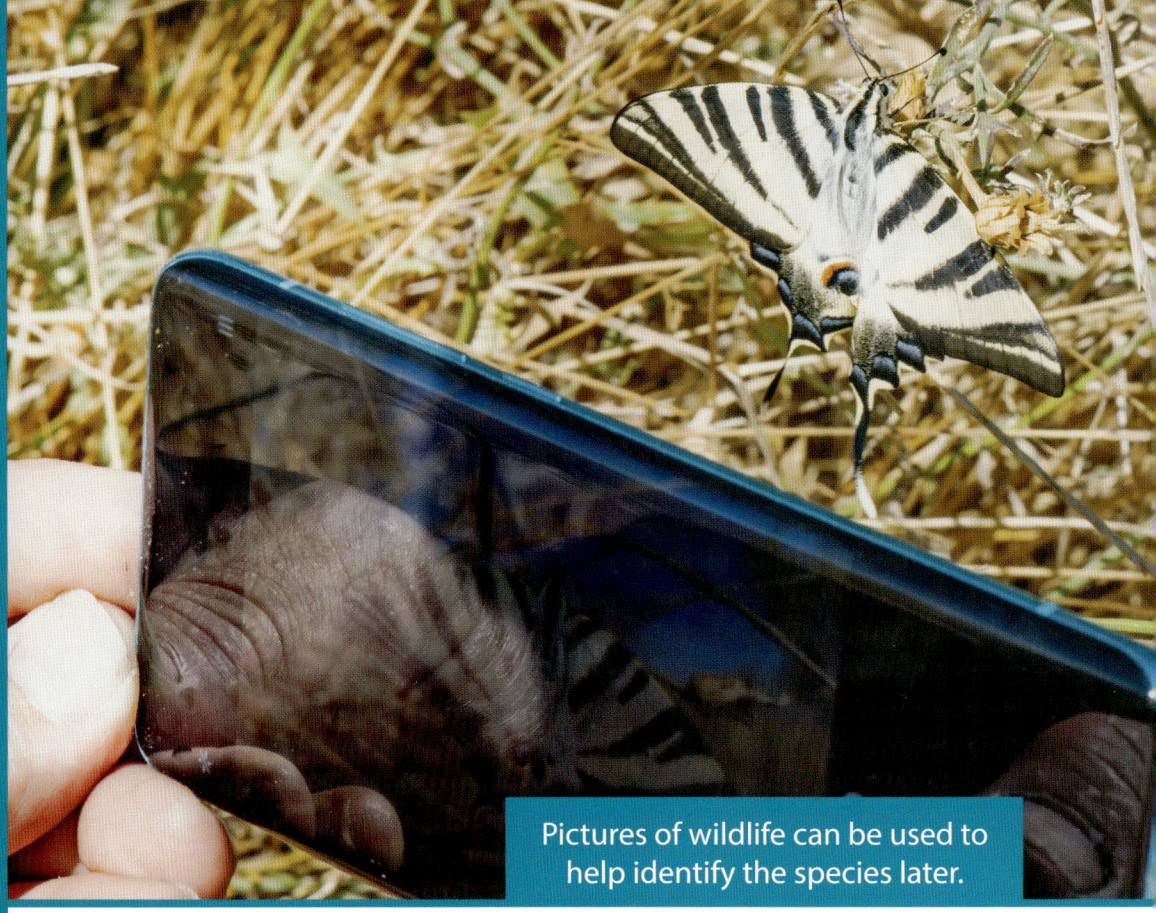

Pictures of wildlife can be used to help identify the species later.

Field guide apps should be downloaded before leaving for a trip. One field guide app that is popular among birders is Merlin Bird ID. This app helps people identify birds by visual characteristics or by sound. The app also lets users upload photos of the birds they see. It compares the photo to its database of birds to help with identification.

For those interested in spotting other animals as well as birds, the iNaturalist app has a database of thousands of species. In addition to being a useful tool for animal identification, it also gives wildlife watchers information about the animals they see. Some national, state, and local parks have their own apps that help visitors identify wildlife.

BINOCULARS

Visual magnification equipment includes binoculars and spotting scopes. While this gear is not necessary for a wildlife watching trip, experienced wildlife watchers may find these tools useful. They allow people to see wildlife from far away. They provide a detailed view of animals. However, powerful magnification equipment can be expensive. It may not be a good fit for casual wildlife watchers or beginners.

Binoculars vary in magnification power.

Binoculars are handheld tools. They have two eyepieces. Most easily fit into a backpack. Other people opt to use a spotting scope. This device sits on top of a tripod, which keeps the scope steady. The scope can swivel on the tripod, allowing users to track wildlife on the move. Unlike binoculars, a spotting scope has only a single eyepiece.

Spotting scopes tend to be less portable than binoculars, but they have high magnification power.

WHAT TO BRING

Binoculars and spotting scopes allow wildlife watchers to view wildlife from far away. These tools also help provide a clear sight of smaller animals. Birders often use spotting scopes to see the details of a bird's feathers. Other wildlife watchers use these tools to see animals while keeping a safe distance from them. These visual aids also make it easier to identify wildlife.

Small binoculars can be a good fit for wildlife watchers who plan to do a lot of hiking during a trip.

Injuries can occur when wildlife watching. Having a first aid kit on hand is helpful in case of emergency.

OTHER ESSENTIALS

A basic first aid kit is another essential for wildlife watching trips. These kits can be assembled at home or purchased. Basic first aid items include bandages, gauze, and antibiotic ointment. Moleskin and other bandages designed to treat blisters should be included. Tweezers and small scissors are also important items in a first aid kit.

WHAT TO BRING

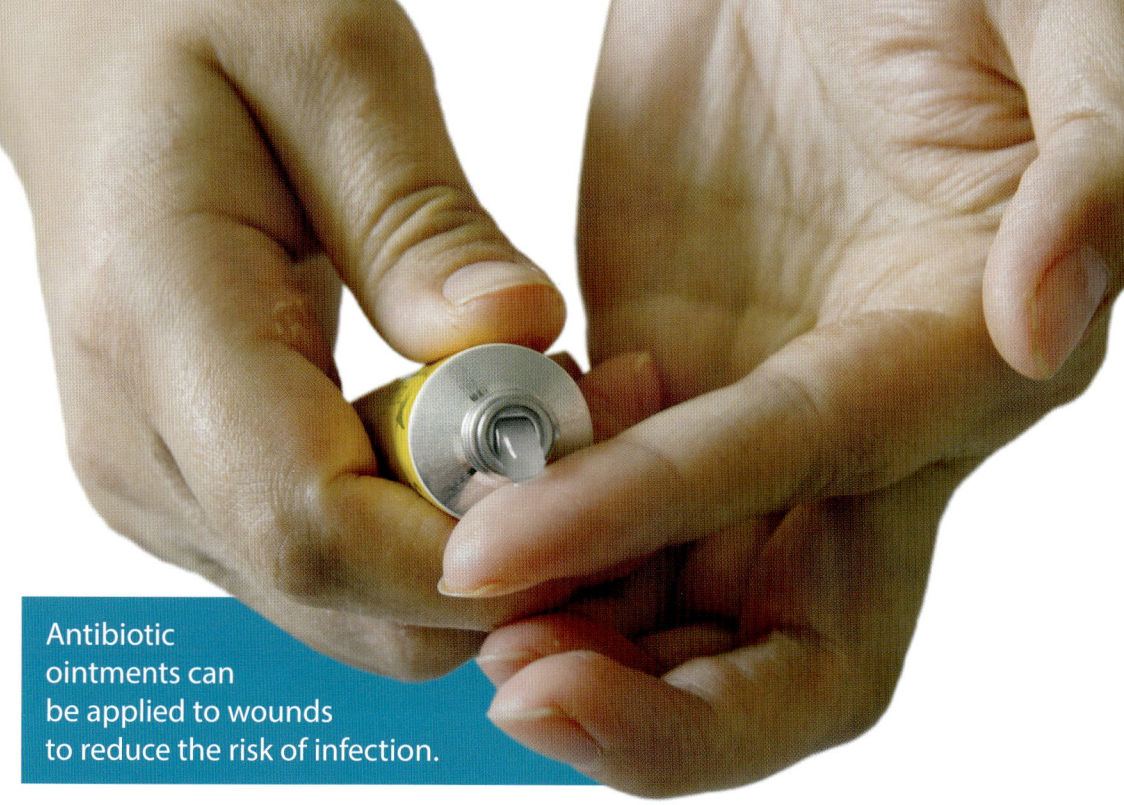

Antibiotic ointments can be applied to wounds to reduce the risk of infection.

Many people carry medication that provides pain relief. The packaging has information about the correct dose for a person's age. Prescription medications should also be included in a first aid kit. For example, people who are allergic to bee stings should carry the medication they need to treat an allergic reaction. Other items in a first aid kit may include hand sanitizer, splints, a thermometer, and an emergency blanket. The entire kit should be stored inside a waterproof bag. People who plan to go on regular wildlife watching trips should take a basic first aid course and a CPR course.

Other safety equipment includes a headlamp, a pocketknife, hand warmers, waterproof matches, and a whistle.

People should also pack and apply sunscreen when spending time outdoors. The sun gives off ultraviolet (UV) rays, which cause sunburn. This is true even on cloudy days. Doctors recommend sunscreen with a sun protection factor (SPF) of 30 for people spending time outdoors. Sunscreen should be applied 15 minutes before sun exposure and reapplied at least every two hours. Sunglasses help prevent UV rays from damaging the eyes. Sunglasses with large rims are most effective at blocking UV rays.

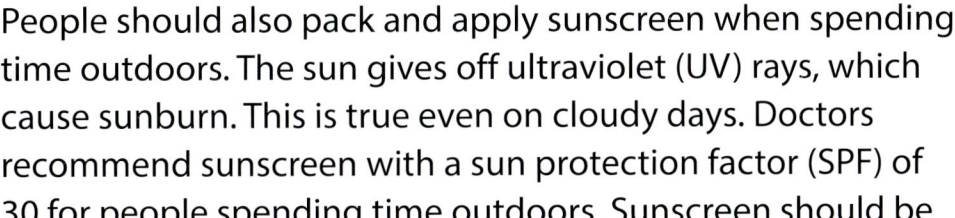

Those who have asthma or other conditions that make breathing difficult should pack an inhaler for wildlife watching trips, especially those that require intense exercise.

WHAT TO BRING

People also need protection from insects during a wildlife watching trip. They should wear long sleeves and long plants to minimize skin exposure. Wildlife watchers may even choose to wear special outerwear to prevent insect bites. Bug net clothing is made of mesh that keeps bugs off the skin. Air can still circulate beneath the netting. People should also pack bug spray to protect against biting insects.

Sunscreen should be applied every hour if people are in the water.

Wildlife watchers can draw the animals they see and take notes of their behaviors in a sketchbook.

   Many people choose to pack a camera so they can take photos during a wildlife watching trip. This could be the camera on a cell phone. Or it might be a separate handheld camera. People should pack extra batteries. Some wildlife watchers choose to bring a journal to record information about the wildlife they see during a trip.

WILDLIFE WATCHING ETIQUETTE

Wildlife watchers must know how to view animals safely and respectfully. They should watch wildlife from a safe distance. The National Park Service (NPS) has distance guidelines for different animals. For small wildlife, they recommend keeping a distance of 50 feet (15 m). That distance applies to squirrels and chipmunks, most birds, and reptiles. For large animals and predators, NPS recommends keeping a distance of 100 feet (30 m). That applies to elk, bighorn sheep, and California condors. Some predators, such as mountain lions and bears, need at least 300 feet (91 m) of space.

Wildlife watching can be safely done from a car.

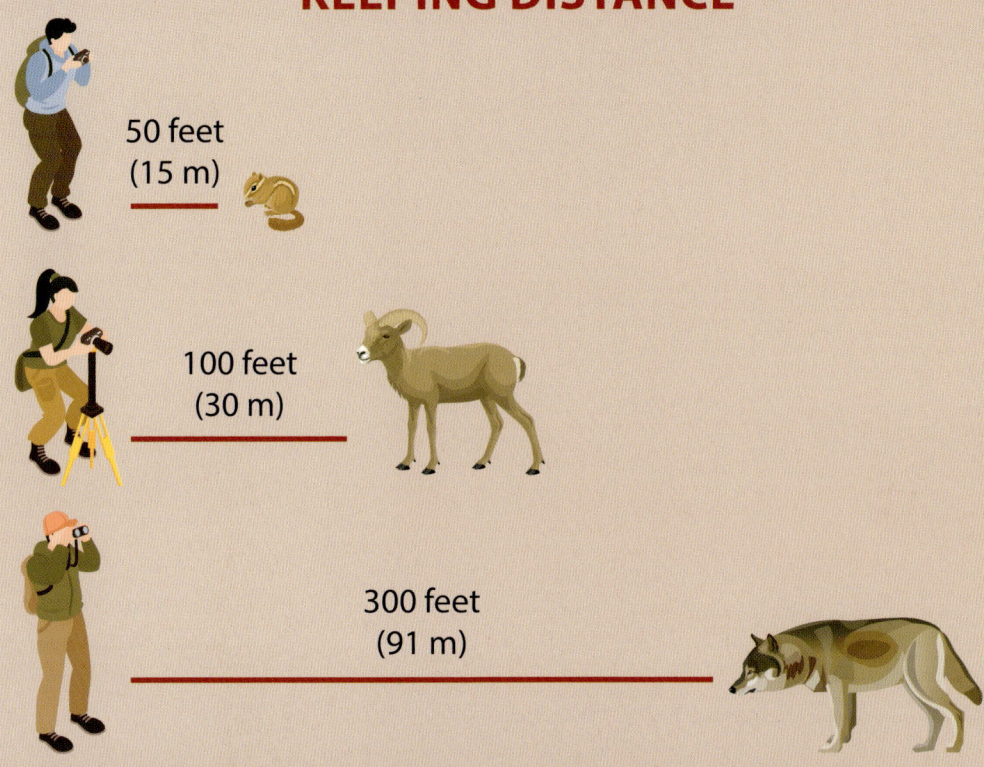

Everyone should maintain a safe distance from wildlife. Large animals and predators should be given more space than small wildlife.

This distance protects both the animals and the viewers. It makes it less likely for animals to become stressed. They continue their activities and behave naturally. If animals do become frightened, the distance gives wildlife room to escape. People should also limit the amount of time they spend watching an individual animal to prevent the creature from getting stressed. Touching wildlife also puts them at risk of catching human diseases.

WILDLIFE WATCHING ETIQUETTE

Wildlife may panic when approached by people. Animals may attack when they are stressed. They can pass diseases to people. Sick and injured animals should be left alone. When concerned about a creature's well-being, wildlife watchers should report the animal to a park ranger or other wildlife authority.

Raccoons can transmit diseases, such as rabies, to people.

Bison cause more injuries in Yellowstone National Park than any other animal, including grizzly bears and wolves.

All animals are dangerous, including small animals and animals that are not predators. In Yellowstone National Park, people are injured or killed every year because they get too close to bison. These animals usually move slowly and generally seem unaffected by visitors. But while bison may seem slow, they can run 35 miles per hour (56 kmh), which is about three times faster than the average person. They can gore people and throw them into the air. Even small animals are unpredictable and can become aggressive.

WILDLIFE WATCHING ETIQUETTE

It is especially important to maintain a safe distance during mating season and when animals are rearing young. People should be on the lookout for nests, dens, and other areas where animals raise their offspring. Disturbing one of these sites may cause parents to flee the area. They may abandon the nest and leave their young behind. This is dangerous for the young animals. They are more vulnerable to predators. Young animals may also be unable to shelter themselves from extreme weather conditions.

If wildlife watchers find an animal they think has been abandoned, they should contact a park ranger or other

Geese are protective of their young and require more space when young are present.

Young animals that have been abandoned require special care in wildlife rehabilitation centers.

wildlife authority. It is not always easy to identify an abandoned animal. Parents may leave their young in a safe location while they hunt for food. People should never approach or touch a wild animal. Young animals such as birds and bison may be rejected by their parents after coming in contact with humans.

LEAVE NO TRACE

In the 1990s, advocates formed the Leave No Trace (LNT) organization to protect natural lands and wildlife. The organization works to educate and empower people to understand the importance of conservation. The original mission of LNT was to protect wild places on land and water. Since then, it has created principles to promote respect for the environment and animals in all outdoor spaces. These principles apply to any outdoor recreational activity, whether in a wilderness area, park, or yard.

WILDLIFE WATCHING ETIQUETTE

Respecting wildlife also means leaving pets at home. In addition to making wildlife watching more difficult, pets cause stress to wildlife. They may chase wild animals. Animals use up energy to escape, which can make survival more difficult in the future. Pets can also cause injuries and death to wild animals. They can pass diseases to wildlife. Similarly, pets are at risk for becoming prey. They may catch diseases from wild animals or be injured by them.

While some hiking trails allow pets, those interested in wildlife watching should leave their pets at home.

Many animals have a reflective layer in their eyes that helps them see in the dark. At night, wildlife watchers can search for flashes of light that may be animal eyes.

Certain noises disturb wildlife and change their behaviors. People may be tempted to use animal and bird calls to locate and attract wildlife. However, using these calls is prohibited in some places. They are a form of harassment because they cause wildlife to change their natural activities. Animals may feel threatened by the recorded calls. They may flee or attack. Lights can also be distracting and stressful to wildlife. When looking for animals that are active at night, people should use as little light as possible. They should let their eyes adjust to the darkness. The moonlight is often bright enough to guide the way.

WILDLIFE WATCHING ETIQUETTE

Parks may have signs reminding visitors not to feed wild animals.

DON'T FEED THE ANIMALS

People should never feed wildlife. It is illegal in many places. Human foods are not healthy for wild animals. They may include nonnatural ingredients. They also introduce food that is not natural to that habitat. Human foods may not supply animals with the nutrients they need. Some foods can also make animals sick.

When animals eat human food, their body functions change. They may get used to human food and become unable to digest natural foods. Wild animals may change their feeding behaviors after eating human foods. They may stop foraging or hunting. Instead, they rely on people for food. Animals that depend on human food lose their natural fear of people. Some animals become bold enough to approach people and beg for food.

Chocolate is toxic to many animals, including birds, raccoons, and coyotes.

WILDLIFE WATCHING ETIQUETTE

Trash must be stored in a secure place when wildlife watching in areas with bears. These animals may rip open trash bags to access food.

DID YOU KNOW?

There is a saying reminding people not to feed the bears: "A fed bear is a dead bear." When bears get used to human food, they lose their fear of people. Wildlife authorities euthanize problem bears.

Wildlife watchers should be mindful of the food they bring in. All food must be stored and cleaned up properly. Animals sometimes steal unattended food. Wildlife may rip tents or backpacks to access human food. Animals that are used to human food may linger near picnic areas and parking lots for meals. This puts them in danger of being hit by a car. Animals that approach humans may need to be put down by wildlife authorities.

POLLUTION

The phrase "pack it in, pack it out" reminds visitors to take all the materials they brought with them when they leave a park or wilderness area. This includes trash and food waste—such as orange peels, apple cores, and nut shells—that can introduce new items into animal diets. Even tiny crumbs should be cleaned up because animals may forage for these small bits of human food.

Volunteering to pick up trash can make natural spaces more enjoyable for people and safer for wildlife.

WILDLIFE WATCHING ETIQUETTE

Animals may mistake trash for food. Plastics can block an animal's digestive tract, leading to starvation. Trash can also puncture internal organs. Plastic pollution has become a serious problem in marine habitats. Every year, millions of animals die because of plastic trash that was improperly discarded. Many marine animals get tangled in plastics, including fishing lines and plastic bags.

Oceans also contain microplastics, which are tiny pieces of plastic smaller than 0.2 inches (0.5 cm). Small marine organisms may eat microplastics. When a larger animal eats that animal, it also consumes the plastic. Studies have shown that ingesting plastic causes tissue damage and harms an animal's ability to reproduce.

Sea turtles may mistake plastic bags for jellyfish, which are a common part of their diets.

Toxic chemicals in sunscreen can contribute to water pollution that causes harmful algae blooms. These algae blooms can make it difficult for wildlife to survive.

Wildlife watchers should also pay attention to the soaps, lotions, sunscreens, and bug sprays they use. These items may include toxic chemicals that can damage ecosystems, harm water quality, and harm wildlife. Products that use toxic chemicals such as oxybenzone should be avoided.

WILDLIFE WATCHING ETIQUETTE

Coral reefs are especially sensitive to toxic chemicals. Those watching wildlife on or near coral reefs must use reef-friendly sunscreens. These sunscreens are made with chemicals such as zinc oxide and titanium dioxide that have less of an impact on the environment. NPS recommends reef-friendly sunscreen even when a trip is not near water to minimize the effect one has on nature. Changes in the water can make it difficult for corals to survive. Harmful chemicals in sunscreen and other products reduce green algae growth. This type of algae is food for many species. When algae levels are low, all marine animals are affected.

The coral reef off the coast of Florida is the largest coral reef system in the continental United States.

European starlings are an invasive species in the United States. They destroy crops and make it harder for native birds to survive.

INVASIVE SPECIES AND DISEASE

Gear must be cleaned before a wildlife watching trip. This prevents the spread of invasive species. Invasive species are species of animals, plants, and other organisms that are not native to an area. They harm the ecosystem. Often, invasive species can reproduce quickly because they have no natural predators. They compete with native species for resources, such as food and habitat. This makes it challenging for native species to survive. Invasive species can negatively impact the biodiversity of an area.

WILDLIFE WATCHING ETIQUETTE

People should wash bikes and other gear before heading into a new region to avoid transporting invasive species.

DID YOU KNOW?

There are more than 6,500 invasive species in the United States.

The spread of invasive species is largely caused by human activity. Mud and seeds can collect in the treads of shoes and bike tires. These seeds can then be carried and deposited in a different habitat. Before leaving an area, all debris needs to be removed from shoes and tires.

Boat hulls can also carry invasive species, such as zebra mussels. These mussels first arrived in the Great Lakes of North America in the 1980s. They were carried by boats from the Caspian and Black Seas in Central Asia. Zebra mussels outcompete native species. They are microscopic in the larval stage. Boaters cannot see whether zebra mussel larvae have attached to their boats. To stop the spread of zebra mussels in the United States, boats need to be properly drained, cleaned, and dried between trips.

Zebra mussels attach to hard surfaces and live in large colonies that can have tens of thousands of individuals.

WILDLIFE WATCHING ETIQUETTE

White-nose syndrome was first identified in bats in 2007. Since then, millions of bats have died from the fungus.

　　Diseases are also spread by human activity. Gear and clothing worn on a wildlife watching trip should be washed to make sure diseases are not brought into a new area. White-nose syndrome is a disease that affects bats. Scientists believe it is partially transmitted by human travel. The disease is caused by a fungus. When exploring caves, visitors may pick up the fungus on backpacks, clothing, or shoes. If these items are not thoroughly cleaned, the fungus then spreads to the next cave they visit. The fungus infects and kills hibernating bats. It can spread among an entire colony of bats.

PROTECTING HABITATS

To protect habitats during a trip, wildlife watchers should stay on roads or established trails whenever possible. If they need to travel off designated trails to relieve themselves, they should walk on durable surfaces. These areas have little or no vegetation and are not harmed by footsteps. Durable surfaces include bare rocks, gravel, sand, snow, and ice.

Staying on trails helps protect habitats and wildlife.

WILDLIFE WATCHING ETIQUETTE

Fragile surfaces are easily damaged by footsteps. The surface may become unstable. The trampled vegetation may not recover. This can lead to erosion and cause nutrient-rich soils to wash away. It is difficult for plants to grow when the

It is possible to walk across thick patches of snow without causing much damage to ecosystems.

STAY ON THE TRAIL

People may be tempted to take shortcuts when hiking on a trail. They may temporarily walk off trail because they want to get to their destination faster. But these shortcuts destroy plants and loosen soil. As vegetation is destroyed, more people may take these unofficial routes. These footpaths erode over time. When water flows across this area, it washes away soil and forms a gully. These gullies continue to erode, degrading the overall habitat.

soil is poor. Erosion may worsen, and the habitat may continue to degrade. This makes it challenging for animals to find food and shelter. Wildlife watchers should avoid traveling on new vegetation, tundra, wetlands, and fragile soils.

WILDLIFE WATCHING ETIQUETTE

Cryptobiotic soils are fragile soils that are often found in the desert. This type of soil is created by microscopic organisms. They release a material that binds the soil together. This creates a crust made of soil and organisms. The crust holds more water than other desert soils. It also keeps moisture from evaporating and prevents erosion. Cryptobiotic soil helps plants grow, creating food sources for animals. But a single footstep can break the crust. This exposes the soil underneath, making it vulnerable to water loss and erosion. It takes a long time for the cryptobiotic soil to recover.

Cryptobiotic soil is darker than surrounding soil and has a spongy texture.

Water holes in the desert are home to unique wildlife species.

In addition, people should avoid walking through puddles, mudholes, and potholes. These fragile surfaces are habitats for tiny organisms. The type of biodiversity in and around a pothole depends on its size and depth. Many of the animals in a pothole cannot be seen without a microscope. But others, such as tadpole shrimp, snails, and diving beetles, may be visible to the naked eye.

WILDLIFE WATCHING ETIQUETTE

Launching from docks helps protect fragile shoreline habitats.

Some shorelines and riverbanks are fragile habitats too. Sensitive dunes should be avoided. People traveling by boat need to plan where to launch their boats and take them out of the water. Docks and designated boat launching areas should be used whenever possible. Rocky beaches are another good choice for putting a boat in the water or taking it out.

CAVE HABITATS

Some people explore caves to see wildlife. They might find bats, frogs, salamanders, crayfish, or fish. Some caves are very isolated. They are home to unique species that are dependent on one another. Even small disturbances can impact the entire cave habitat. People should explore caves only with a knowledgeable guide. This protects the wildlife and prevents people from getting lost.

Proper disposal of human waste also protects habitats. Improper disposal leads to soil and water pollution. It can spread disease and make outdoor experiences less enjoyable for others. Parks where the soil is dry, rocky, or frozen may require people to pack out solid human waste. That's because these conditions make it difficult for waste to decompose.

Some parks may have restrooms for visitors to use.

WILDLIFE WATCHING ETIQUETTE

In most places, burying waste is an accepted method of disposal. People first dig a cathole. The best places to dig are sunny spots with loose, dark, rich soil. The sun and soil help waste decompose quickly. Catholes should be about 6 to 8 inches (15 to 20 cm) deep and 4 to 6 inches (10 to 15 cm) wide. They must be located at least 200 feet (61 m) away from trails and water sources. Toilet paper may be buried, but some parks and wilderness areas require toilet paper to be packed out. All feminine hygiene products must be packed out as well. After use, catholes should be filled with the original dirt and camouflaged with natural materials.

Wildlife watchers should check to see if there are special rules about waste disposal in the park or region they are visiting.

Wildlife watchers on an ocean or a river should plan to pack out waste if they do not have easy access to land. There are exceptions to solid waste disposal in remote ocean locations. In some areas with active waves and currents, waste may be left directly in the ocean. Before a trip, people should check the regulations in the area.

Wildlife watchers who are planning trips that require relieving themselves outdoors should pack a small shovel to bury waste.

DID YOU KNOW?

Urine has little effect on plant and soil health. But the salts in urine can attract wildlife. People should rinse the area with water to minimize this risk.

WILDLIFE WATCHING ETIQUETTE

Protecting habitats includes leaving all natural materials where they are found. People should not take seashells, rocks, sticks, wildflowers, or feathers home with them. These items are part of the habitat. Beachgoers may want to collect seashells during a visit. But these shells are important to many animals.

Birds may use shells as nesting material. Shells may wash back out to sea, providing hiding places for small fish. Hermit crabs change shells as they grow. Shells that remain on shore help prevent beach erosion. They provide food for small organisms. Over time, these shells break down and become sand.

Seashells are important parts of beach habitats.

WILDLIFE WATCHING ETIQUETTE

Hummingbirds and other pollinators rely on wildflowers for food.

Likewise, picking wildflowers harms habitats. Many wildflowers reproduce slowly and take many years to mature. If every visitor to an area picked a wildflower, it would be difficult for the ecosystem to recover. In some areas, it is illegal to pick wildflowers. Wildflowers support pollinators, such as insects and hummingbirds. These animals drink nectar from flowers. Some pollinators have a small home range or a very specific diet. The loss of flowers in their ranges can make it challenging to find the food they need to survive.

RESPECTING OTHER WILDLIFE WATCHERS

Wildlife watchers should respect other people who want to enjoy the outdoors. Group sizes should be kept small. Everyone should talk quietly. Respecting others includes being mindful of the use of technology, especially music. Loud music may scare away wildlife. People who want to listen to music should use earbuds instead of speakers. Respect for others is another reason to keep pets at home. Wildlife watchers should take steps to make the outdoors a safe space for wild animals and an enjoyable place for other viewers.

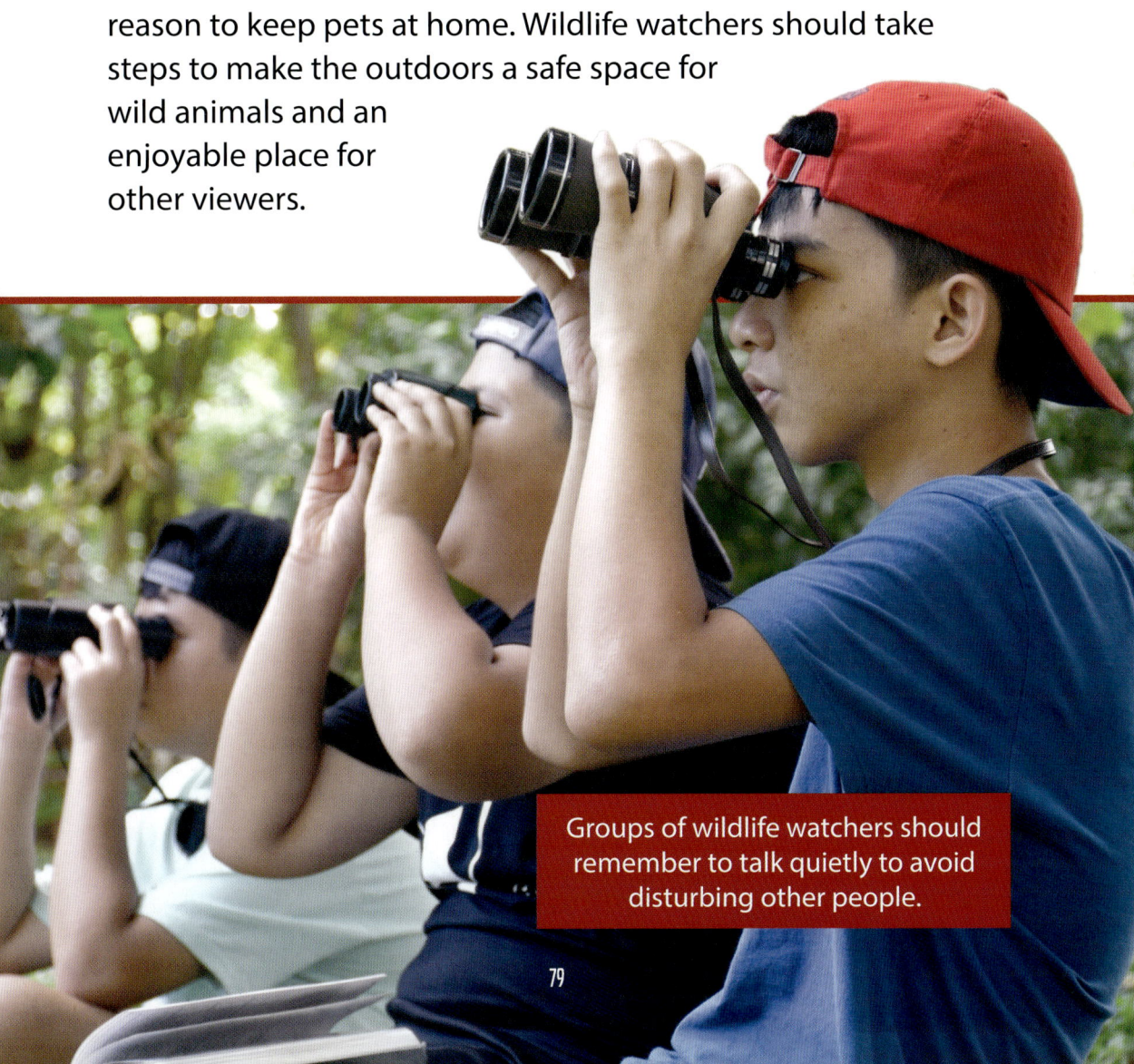

Groups of wildlife watchers should remember to talk quietly to avoid disturbing other people.

TIME AND LOCATION

Wildlife watchers must consider time and location when choosing a destination for a wildlife watching trip. Some animals are found only in specific parts of the country. They may live only in certain types of habitat. Animals may be less active depending on the time of day or part of the year. Wildlife watchers may want to see a specific animal. They should consider how these factors influence the target species. They may also need to consider whether migration patterns and mating seasons affect the likelihood of seeing the animal.

Reptiles, such as snakes, are easier to spot when the weather is warm. These animals are much less active during the winter.

Sea lions are found only in the Pacific Ocean, so wildlife watchers in the United States will need to go to the West Coast to spot these animals.

LOCATION

Wildlife watchers who are in search of specific animals may need to travel to certain areas of the United States. Research informs people where the target species is most likely to be seen. For example, American alligators live in the southeastern part of the country. People need to visit Alaska to see caribou. Large numbers of wild bison are found in Yellowstone National Park, which lies mostly in Wyoming.

TIME AND LOCATION

Forests, mountains, deserts, wetlands, prairies, oceans, streams, and lakes are all types of habitats. Each one supports different species. Wildlife watchers should know what habitat a target animal lives in. For example, moose and puffins both live in Maine. But these animals live in different habitats. Seeing moose requires heading inland to forested areas. Puffins roost along the rocky Atlantic coast or on islands.

People can further narrow their search within a habitat. To see pikas in the mountains, wildlife watchers need to go above the tree line to boulder fields or high alpine meadows. And people who want a chance to spot a greater sage-grouse need to visit grasslands with sagebrush.

Pikas store wildflowers and grasses in their dens so that they have food during the winter.

Great blue herons can be found in edge habitats. They live in freshwater and saltwater habitats and also search for food in grasslands.

Edge habitats are the transition areas where different habitats come together. Examples of edge habitats include regions where forests meet meadows, or the shorelines of rivers and lakes. Human habitats, such as a neighborhood, a farmer's field, or a road, also form edge habitats. Animals found in edge habitats are adapted to live in multiple habitats. For example, elk graze in open meadows at the edges of forests. They use the forest to hide. For wildlife watchers, these zones allow them to see a range of animals from different habitats.

TIME AND LOCATION

Wildlife bridges help animals safely cross busy roads.

Wildlife corridors provide excellent wildlife watching opportunities too. These are areas that animals use regularly for migration or to find food. Some corridors are natural, such as the migration routes caribou use each year. Other wildlife corridors are human-made. Human development can block natural pathways. Structures can be built over busy roadways to connect natural habitats and allow wildlife to cross safely.

WATCHING BATS IN AUSTIN

Human development has had a negative impact on wildlife, but some species have adapted to life in urban spaces. More than one million Brazilian free-tailed bats roost underneath a bridge in Austin, Texas. The space under the bridge is warm, dark, and safe. Wildlife watchers take boat tours to view the bats between mid-March and mid-October every year. The colony of bats emerges about 30 minutes before sunset and swarms the skies. The best places to see them are on the bridge or on the river below. People also take boat tours to view the bats or paddle on the river in kayaks.

Fish ladders are another type of human-made wildlife corridor. They allow fish to travel over dams to reach their breeding grounds. For example, Chinook salmon spend most of their lives in the Pacific Ocean. They breed in freshwater rivers. Fish ladders help the salmon cross over dams. Wildlife corridors offer unique wildlife watching opportunities since they are routes animals regularly use.

TIME AND LOCATION

TIME OF DAY

Selecting the time of day is part of planning a successful wildlife watching trip. Different animals are active at different times of the day. Some animals are diurnal. This means they are most active during the daytime. They are less active at night and may hide, rest, or sleep during this time. Being active during the day makes it easier for diurnal animals to see prey. Diurnal animals include moose, bald eagles, groundhogs, and painted turtles. People who want a chance to see these animals should be at their destinations during the middle of the day.

Painted turtles bask in the sunlight and are most active in the morning.

Sea otters are diurnal animals that can be seen eating and swimming along parts of the Pacific Coast.

TIME AND LOCATION

Nocturnal animals are active at night. They spend the daylight hours resting. In the deserts of the Southwest, many animals are nocturnal in order to avoid the heat of the day. By staying inactive during this time, these animals keep cool and conserve water. Kangaroo rats are an example of a nocturnal desert animal. Wildlife watchers are most likely to see one of these creatures after the sun has set. The darkness also helps protect animals from predators. It is more difficult for some animals to locate prey in the dark. But some predators hunt at night because that is when their prey is active. Owls and big cats such as Florida panthers are adapted to hunt in the dark. They have excellent hearing and night vision.

Kangaroo rats sleep in underground burrows during the day.

Barn owls can see very well in the dark. They also have sensitive hearing, which allows them to hunt at night.

TIDE POOLS

Tide pools are located along rocky coastlines of the Pacific Northwest and New England. They are accessible only at certain times of the day. Tides constantly ebb and flow. Tide pools disappear at high tide and reappear at low tide. Tide charts assist wildlife watchers who want to see ocean wildlife up close. These charts have information about the timing of high and low tides.

TIME AND LOCATION

Crepuscular animals are active at dawn and dusk. That makes sunrise and sunset the best times to see them. These times of day are cooler than the middle of the day. There is enough light to find food. The low light also makes them less visible to predators. Crepuscular animals include porcupines, deer, coyotes, and many songbird species. Checking with wildlife experts in an area helps wildlife watchers determine the best time to see certain animals.

Porcupines can climb trees. They eat many types of plants, including the needles of conifer trees.

American bullfrogs and other amphibians tend to be more active in wet conditions.

WEATHER

Animals behave differently depending on the weather and season. Conditions such as wind, precipitation, and temperature can create challenges for wildlife watchers. They also influence animal behavior. Knowing the habits of animals helps wildlife watchers find and locate a specific species.

TIME AND LOCATION

> Because of the shape of their wings and tails, kestrels are able to hover in place in strong winds.

Wind affects an animal's ability to hear. Prey animals are more alert as they watch for predators. Insects are less active on windy days because strong winds make it difficult for these small creatures to move around. Birds and other animals that feed on insects are also less active when there are strong winds since it is more difficult to locate their prey. Most birds take shelter when it is windy. Some wait in tree cavities. Others perch on a branch close to the trunk of a tree, out of the wind. Wading birds shelter among vegetation. They find a safe spot to wait until the weather passes.

Rain affects animal behavior too. Many animal species become active just before a storm. They know when a storm is coming because they sense a change in air pressure. Insects increase their activity before it rains. This creates an opportunity for insect-eating animals to feed. Before it rains, many bird species look for insects. Because animals are more active before a storm, this can be a good time to watch wildlife.

Most insects take shelter when it is raining. They may hide under flowers, leaves, fallen logs, and other objects.

TIME AND LOCATION

Once the rain begins, most animals seek shelter. The rain interferes with their senses, making it difficult for them to see, hear, and smell. Rain can also cause challenges in maintaining body temperature. Insects are unable to fly in the rain, so they take cover. As a result, animals that prey on insects also find shelter to conserve their energy. For these reasons, spotting wildlife may be more difficult during rainy weather. But animals will eventually need to emerge to find food if the rain continues for a long time.

Despite their small size, hummingbirds are able to fly in the rain.

Rabbits and many other animals eat plants and depend on rainfall to help grow their food sources.

After the rain stops, wildlife emerges to forage and hunt. This is a good opportunity for wildlife watching. Abundant plant life supports insect populations. Plants and insects are food sources for many animals and can attract more wildlife.

TIME AND LOCATION

Squirrels bury nuts throughout the year so they will have food to eat in the winter.

As with rain, just prior to a snowstorm is a good time to watch wildlife. Many animals are busy feeding and preparing for the storm. But fallen snow creates challenges for wildlife. Snow covers food sources such as nuts, seeds, and plants. It is difficult for animals to move through deep snow. Deer move into forests where the snow is less deep. Some species survive under the snow without coming to the surface. Because of this, areas covered in deep snow may have little wildlife to view.

In areas where snowfall is light, animals tend to gather in areas that face south. In these areas, the snow melts faster than in other areas because they get more direct sunlight. Elk, deer, and other large mammals may avoid areas of deep snow as they seek food sources. Even without snow, animals behave differently in cold weather. Smaller animals burrow underground to wait out the cold.

In places that experience all four seasons, animals prepare for winter when the temperature drops in the fall. Some begin their migrations south to warmer climates. Others begin storing food because food is scarce during the winter. Visitors to the mountains may see pikas gathering plant materials. They take this food to their underground burrows to store for the winter. Moles may be seen collecting worms before the ground freezes. They store the worms in aboveground spots.

Geese begin their migration to the southern United States in the fall. They typically travel in family groups.

TIME AND LOCATION

As many as 74 grizzly bears have been seen hunting for salmon on the McNeil River at one time.

    Some animals prepare for hibernation during the fall. These animals need to eat a lot of food to prepare for the cold winter months. As they eat, they put on a layer of fat. This fat provides them with energy that helps them survive the winter. Grizzly bears are hibernating animals. Wildlife watchers may travel to Alaska in the fall to watch these bears gather at streams and rivers to hunt salmon. One of the largest gatherings of grizzly bears occurs at the McNeil River in Alaska each year. Once winter arrives, hibernating animals spend the months in dens or burrows. They become active again in the spring.

    Reptiles, amphibians, and most fish are cold-blooded, which means they do not produce their own body heat. Instead, they rely on the temperature of the air or water around them to regulate their body temperatures. Because of this, cold-blooded animals seek shelter when the weather cools. For example, fish, turtles, and frogs retreat to the bottoms of lakes and ponds where the water is warmer. Snakes move into dens. Other animals dig under leaf litter on the forest floor for insulation. Cold-blooded animals move slowly in the winter. They need to conserve energy. Seeing these animals on cold days is unlikely. However, unlike hibernating animals, cold-blooded animals may emerge or move around on warm winter days to bask in the sun. This behavior is called sunning.

Catfish move to deeper areas of the water during the winter.

TIME AND LOCATION

As the weather warms in the spring, some animals migrate back to their summer ranges. Warmer weather also brings many animals out of hibernation. They emerge to find food and to look for mates. Insects and cold-blooded animals also emerge and become more active as the weather warms.

As cold-blooded animals, Texas horned lizards depend on the sun for warmth. They are more active during the summer.

Deer typically give birth in the late spring and early summer.

Many animals are active during the summer. Some raise their young during this time of year. They need to spend a lot of time foraging and hunting to provide enough food for their offspring. Because wildlife is so active and abundant, summer is one of the best seasons to plan a wildlife watching trip.

TIME AND LOCATION

Wild horses rely on watering holes as water sources and places to cool down.

During long periods of warm, dry weather, some animals gather around water sources to cool off. Birds are among the animals that use ponds, lakes, and other water sources to cool themselves. Even butterflies sometimes gather near shallow water. Deer and other large animals may wade into the water to cool off. Wildlife watchers should look for these water sources when searching for animals in hot weather.

At high temperatures, animal activity may slow down. Many animals seek shelter to stay cool. If the heat is ongoing, animals may shift their behaviors so they are more active during cooler parts of the day. For example, diurnal animals may adopt a crepuscular or nocturnal schedule because it is cooler when the sun is not directly overhead.

Rattlesnakes may be seen at any time of the day. However, these snakes are most active around dawn and dusk when temperatures are cooler.

MATING AND BREEDING SEASONS

Animals mate and breed at different times throughout the year. Some animals have mating rituals that people enjoy watching. For example, wildlife watchers may head to Rocky Mountain National Park in Colorado to see elk. During autumn, herds of elk often gather in open meadows. Male elk are called bulls. The bulls compete to mate with females. The males use their antlers to fight one another and prove who is the strongest and healthiest. Bulls also make bugling calls that echo off the mountains. Some calls are used to communicate with females in the bull's group. Other calls warn rival males to stay away.

An elk's bugle can be heard up to 0.75 miles (1.2 km) away.

Elephant seals are named for their large size and long noses.

In California, December and January are the best months to see elephant seals battling for a mate. Male elephant seals can weigh as much as 2.5 tons (2.3 metric tons). The males bellow and honk as they fight each other on the beaches. The strongest males win the right to mate.

DID YOU KNOW?

People interested in watching tarantulas can plan a trip to southwestern Colorado. In September, these spiders roam across the grasslands in search of mates. The best time to see tarantulas is an hour before sunset on warm days when the winds are calm.

TIME AND LOCATION

A male greater sage-grouse, *left*, performs its mating dance for a female sage-grouse, *right*.

Many birds build nests, mate, and reproduce in the spring, making this season an excellent time for bird-watching. Between March and May, people can watch the mating ritual of the greater sage-grouse in the western United States. Males puff out their chests and spread their tails. They inflate and deflate the yellow air sacs on their chests. Females choose one male to mate with based on the performance.

Many amphibians are found in groups in or near ponds or other water sources during the spring. They often return to the place where they hatched to find a mate. In some areas, hundreds of amphibians return at the same time, providing a great opportunity for wildlife watchers.

Most animals give birth in the spring. The warmer weather makes it easier for young animals to maintain their body temperature. Food is abundant in the spring, and parents can provide for their young. For wildlife viewers, it is the time of year to see young animals. However, people need to keep a safe distance from animals. They should never get between parents and their young.

Male spring peepers make a peeping noise when they are ready to mate. Their vocal sacs inflate as they produce this sound.

TIME AND LOCATION

Birds can be aggressive during nesting season. Some hiss or make sharp vocalizations. Other birds try to make themselves look bigger by puffing out their feathers. Some birds dive-bomb people or animals they view as threats.

Red-winged blackbirds are extremely territorial during breeding season. They will attack hawks and other large birds.

Gray whales may swim more than 12,000 miles (19,300 km) each year as they migrate in the Pacific Ocean.

MIGRATION PATTERNS

Many wild animals migrate. Some migrate south during the winter to shelter in warmer climates. They return north to cooler climates for the summer. Other animals make seasonal migrations to locations where there is more food available. Some animals migrate to safer locations to breed.

TIME AND LOCATION

Caribou migrate each year to breed and escape predators.

Migration paths can be over land, through the air, or along waterways. Some wildlife travel extremely long distances each year, such as caribou in Alaska. Caribou herds travel more than 2,000 miles (3,220 km) between their summer and winter ranges. Hummingbirds migrate between locations in North America and Mexico. Some travel more than 1,000 miles (1,610 km). Some salmon species migrate hundreds of miles. Their migration begins in the ocean, and the fish swim upstream through rivers to reach their spawning grounds. Understanding migration patterns informs wildlife watchers of the best times and places to see certain species.

THE MISSISSIPPI FLYWAY

The Mississippi Flyway is an aerial migration route that follows the Mississippi River. It stretches from the river's headwaters in Minnesota to the Gulf of Mexico. More than 300 bird species use this critical migration route. They travel north in the spring. This is where their breeding and nesting grounds are located. In the fall, they return south to their wintering grounds on the Gulf Coast. Some continue beyond the flyway to parts of Central or South America. The Mississippi Flyway is a great location to watch migrating birds.

FINDING AND IDENTIFYING ANIMALS

For a successful trip, wildlife watchers need to know how to find and identify wildlife. It is helpful to think like an animal. People should consider what animals need to survive, their habits, what they eat, and where they take shelter.

Patience is important when wildlife watching.

Staying still and quiet increases the chance of seeing wildlife.

    Finding wildlife involves staying quiet and hidden, being alert, engaging all senses, and patience. It also requires learning how to identify the signs and sounds of animals. When looking for wildlife, people should talk only when necessary. Cell phones, cameras, and other digital devices should be silenced. Loud, sudden, and unnatural noises can scare away wildlife. Noisy gear and rustling leaves and branches may also startle animals, who may run away or hide. If possible, wildlife watchers should find a place to sit. If people are quiet, animals are more likely to enter the area and behave normally.

FINDING AND IDENTIFYING ANIMALS

Many animals, including opossums, have a strong sense of smell. Avoid putting on perfume or other scented products in order to blend in with the surroundings and increase chances of spotting wildlife.

When looking for animals, wildlife watchers should try to stay camouflaged. People can also hide behind natural materials in the habitat. They should stay downwind of target animals. Many animals have an excellent sense of smell. They can detect humans long before people see them. Before going out in the field, scented products such as perfumes, colognes, laundry detergents, soaps, and lotions should be avoided.

TRACKS

People can look for signs of wildlife to help locate animals. Animal tracks, or footprints, are one sign to look for. Each species has a unique set of tracks. Wildlife watchers can examine tracks to figure out what type of animal made them. However, wildlife watchers should be aware that animals in the same family may have similar tracks. For example, white-tailed deer, elk, and moose are all members of the deer family. Their tracks have many similarities. Features such as shape, size, and distance between individual tracks help people identify the animal tracks.

Wildlife watchers can learn to identify animals from their tracks.

People can look for certain details when studying the shape of animal tracks. They look to see if the footprints have fingers or toes. They can also look for visible claws. Some animals have hooves. Hoofprints look like two halves of an upside-down heart. Wildlife watchers can compare front prints with hind prints. Some animals may have differently shaped front feet and hind feet.

Mammal tracks come in many shapes. Rodents, including chipmunks, muskrats, and squirrels, have four paws. Their front paws have four toes, and their back paws have five toes. Both sets of prints show claws. Animals that are part of the weasel family, as well as opossums, bears, and beavers, have five toes on both the front and back feet. Animals in the deer family leave hoofed prints with two toes. Canines, such as foxes and

Beavers have distinct tracks. Their back feet have webbing and are larger than their front feet.

COYOTE TRACKS VS. MOUNTAIN LION TRACKS

COYOTE TRACK
Four toes
Visible claws
About 2 inches
(5.1 cm) long

MOUNTAIN LION TRACK
Four toes
No claws
About 3 inches
(7.6 cm) long

People may confuse coyote and mountain lion tracks. However, mountain lion tracks do not have visible claw marks. They also tend to have larger feet than coyotes.

wolves, have four toes on their front and hind feet. Canine tracks also include a heel pad behind the toes. Their tracks leave visible claw marks. Felines include mountain lions and bobcats. These animals have tracks that are a similar shape to canine tracks. Felines have four toes and heel pads. Though felines have claws, they are not visible in feline tracks. Their claws are retractable. They tuck under the skin when not in use. Felines do not need to use their claws to walk, so their tracks do not show claw marks.

Bird tracks tend to be narrower than mammal tracks. Many birds have four thin toes. Typically, three of these toes point forward and one points backward. Songbirds, doves, ravens, and many more bird species leave tracks of this shape. Owls and woodpeckers have different tracks. They have two toes that point forward and two that point backward. Gulls, ducks, terns, and geese have webbed feet. They have skin in between each of their three toes. The toes and webbing are visible in their tracks.

Tracks on the sand may get washed away by water.

DEER TRACKS VS. MOOSE TRACKS

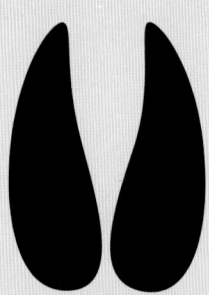

DEER TRACK
No larger than 4 inches
(10.2 cm) long
Heart-shaped

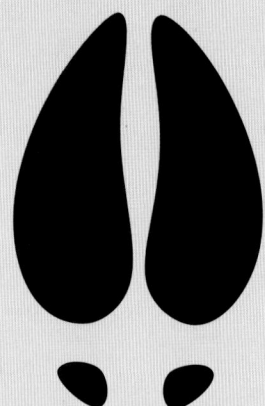

MOOSE TRACK
Typically 4.5 to 6 inches
(11.4 to 15.2 cm) long
Tracks narrow at the top

Moose tracks are much larger than deer tracks. They also have a slightly different shape.

The size of tracks is another factor that helps with animal identification. Since moose and mule deer are members of the same family, their tracks have similar shapes. But moose prints are much larger than mule deer prints. Members of the feline family have similar tracks too. A mountain lion print is roughly 3.5 inches (8.9 cm) wide, whereas a bobcat print is only about 2 inches (5.1 cm) wide. Geese and ducks also leave prints that look alike, but a goose print is 1 to 1.5 inches (2.5 to 3.8 cm) wider than a duck print.

The pattern of the footprints is another clue when identifying tracks. Some animals waddle. Others hop or bound.

FINDING AND IDENTIFYING ANIMALS

Some animal tracks leave a zigzag pattern. Coyotes, deer, and foxes are zigzaggers. When they walk, the foot of the back leg lands in or near the print of the front foot, leaving a pattern that weaves back and forth.

Habitat and season should also be considered when using tracks to identify an animal. Some animals are found only in

Rabbits and other animals that hop leave behind a unique pattern of tracks.

Martens have sharp claws that help them climb trees.

certain types of habitats. They may only be in an area at certain times of the year. For example, minks and martens leave prints that are similar in size and shape. But mink tracks are more common near water, while martens live in forested areas.

If tracks are fresh, the animal that made them may still be nearby. Wildlife watchers can look at the edges of the track to help determine how fresh the track is. Tracks with sharp, well-defined edges are fresh. Factors such as temperature, rain, sun, and wind affect tracks. These factors can cause a track to erode or dry out. The edges become less defined. People can imprint their own track next to the animal track to compare the

the freshness of the tracks. If the tracks appear fresh, observers can carefully follow the tracks. The tracks may lead to a den, resting place, or feeding location. These places can allow wildlife watchers to get a close look at animal behaviors. People should always stay alert and minimize disturbance of animals.

Checking the moisture around the edges of a track helps determine how fresh it is.

SCAT

Scat is another sign that animals are in an area. People should never touch scat. The size, shape, and consistency of scat provides information about the animal that left it. As with tracks, scat is similar among animals of the same family. Many hoofed animals leave pellet-like scat. Larger animals leave larger pellets.

Hares and rabbits have scat that looks similar to that of hoofed animals, but their scat is slightly rounder. Thick, cord-like scat belongs to raccoons, canines, and bears. Reptile droppings are long and thin cords that often have white ends. Among wild turkeys, it is even possible to determine the sex of the bird by its scat. Females leave scat that is spiral-shaped. Males' scat is J-shaped.

Animal scat can carry diseases, so it is important not to touch it.

Because coyotes primarily eat meat, their scat may have bones or hair in it.

The content of scat provides further information about the animal that left it. Coyotes hunt small animals and also eat fruits and other vegetation. During the winter, their diets include more meat. While coyote scat is long and tubular like that of pet dogs, it may contain fragments of bone as well as bits of fur. Black bear scat comes in many shapes and varies depending on the bear's diet. In the late summer and fall, bears mainly eat fruits and berries. This diet causes their scat to look like a large round blob, which can look similar to bison scat.

But bear scat often has remnants of seeds. Bison have a different diet and eat mostly grasses, which do not have seeds.

The visibility of scat is another clue that aids wildlife watchers in identification. Some animals leave their scat in the open, while others bury it. Coyotes leave scat in prominent places, such as trail crossings, to mark their territory.

Armadillos are known to bury their scat.

Deer have pellet-shaped scat because of their diets and the shape of their digestive tracts.

Felines often cover their scat. Raccoons return to the same spot over time to relieve themselves.

Habitat can also help identify scat. While many hoofed animals have scat that looks similar, wildlife watchers should consider which species would be in the particular habitat. Mountain goats and pronghorn leave scat pellets that are roughly the same size. But these animals are found in different habitats. Mountain goats live in

FINDING AND IDENTIFYING ANIMALS

Bear scat tends to have a loose shape in the spring and early summer when their diet consists mainly of plants.

steep, craggy mountains, whereas pronghorn are found on the prairie.

When wildlife watchers come across scat, it is necessary to note how fresh it is. Fresh scat looks moist and often has a shiny appearance. Older scat looks dry and dull. It loses its shape and falls apart. If it hasn't rained and the scat is still moist, it was left recently. The animal may still be nearby.

OTHER SIGNS

Aside from tracks and scat, there are other signs that animals may have recently traveled through an area. In places with ground cover or brush, grasses and plants may be flattened or trampled. It may look like a trail is running through vegetation. Low-hanging branches may be broken. Soil, rocks, and sand may look disturbed. These signs indicate travel routes for animals or places where animals may have slept.

Snakes shed their skin as they grow. A snakeskin is a sign that there may be a snake nearby.

FINDING AND IDENTIFYING ANIMALS

    Burrows are one place where animals live and rest. Animals dig burrows underground for shelter. These holes and tunnels vary in size and shape. Larger burrows are likely inhabited by larger animals. Armadillos make round holes that are 6 to 7 inches (15 to 18 cm) wide, whereas chipmunk holes are approximately 2 to 3 inches (5 to 8 cm) wide. Many animals make round burrows, but desert tortoises make burrows that are wide and short. They are similar to the shape of the tortoise's shell.

> Prairie dogs live together in colonies. If there are several tunnels or holes in an area, the burrows may belong to prairie dogs.

A red fox typically gives birth to four or five pups each year.

Some animals take shelter in dens. Dens can be in a tree hollow in or underneath a tree. They may be located in a hillside or among rocks. Red foxes use dens during the breeding season to raise their pups. The female stays in or near the den before giving birth. The pups remain in the den for four to five weeks. The family may use the same den each year. Most birds build nests for shelter and to raise their young in. Birders should be on the lookout for nests, feathers, and broken shells.

FINDING AND IDENTIFYING ANIMALS

Food remains and feeding sites are also signs of wildlife. Vegetation and fungi may be nibbled. There may be leftovers on the ground. For example, squirrels eat pine cones. They pull off the scales of the pine cone to get to the seeds at the core. The scales and core of the pine cone are left behind as scraps. Rocks may be overturned. Rotting logs could be torn open. Other signs of feeding include bones and animal remains. People should never touch animal remains. They should stay alert for predators that may still be in the area.

Animal carcasses can attract wildlife such as turkey vultures, which eat dead animals.

Grizzly bears mark trees with their scent as they rub against the bark.

    Another sign of wildlife is damage to tree bark. Moose, deer, and elk rub their antlers on trees. These animals grow new antlers each year. They are covered in velvet that supplies the antlers with oxygen and blood, which helps them grow. The velvet dries once the antlers have stopped growing for the season. The antlers becomes itchy. The rubbing removes the velvet and relieves the itch. It also leaves marks on the trees.

    Black bears also use trees for itch relief. They rub their backs on tree trunks. Fur may catch on the bark. Bears also leave claw marks on trees as they climb. Some bears even pull bark off trees to eat or as nesting material.

FINDING AND IDENTIFYING ANIMALS

Wildlife watchers should also be on the lookout for the animals themselves. Most animals are well camouflaged in their natural habitats. They may be difficult to see. Often only part of the animal is visible. People should look for patterns in forest landscapes. For example, there are few branches close to the ground. Tree trunks and vegetation close to the forest floor run vertically. People can scan the undergrowth for horizontal lines. These may be the outline of an animal. Rustling leaves, moving branches on a nearby bush, falling feathers, or rolling rocks

Some animals, such as the common walking stick, are experts in camouflage. Wildlife watchers need to pay close attention to spot these animals.

The American anole can camouflage with its surroundings. This can make it difficult for wildlife watchers to spot one.

are other signs that may indicate wildlife is nearby. Splashes, ripples, and bubbles on or near water may indicate that an animal is below the surface.

FINDING AND IDENTIFYING ANIMALS

Wildlife can be located by listening for animal sounds and watching for movement.

CALLS AND SOUNDS

Along with looking for signs of animals, wildlife watchers need to listen to their surroundings. Rustling leaves or snapping twigs may reveal an animal moving through the area. Fish and other aquatic animals splash. Animals also make many vocalizations. Smaller mammals squeak. Frogs croak. Foxes and coyotes bark or howl. Each sound an animal makes has a different meaning. Calls may be used to attract mates, while other noises are made to defend territory or to warn other animals to stay away.

Mountain lions make a variety of sounds. One noise that they make sounds like a person whistling or a bird chirping. A mother mountain lion makes this sound to communicate to her young. Female mountain lions also make mating calls. These calls sound like high-pitched screams. Mountain lions also hiss and growl when threatened.

> Mountain lions do not have an established mating season. People can hear a mountain lion scream at any time of the year.

FINDING AND IDENTIFYING ANIMALS

The sidewinder is one of 15 species of rattlesnakes that can be found in the United States.

The pig frog lives in the southeastern United States.

As with many animals, frogs use sound to attract mates, mark territory, and warn others of predators. Each frog species makes a different sound. A bullfrog sounds like a horn. The pig frog sounds like a grunting pig. And a barking tree frog makes a sound similar to a barking dog. Knowing the sounds helps people locate and identify different frogs. This skill takes time and practice to develop.

Similarly, understanding the different sounds birds make is important for birders. Most birds have a song and a call. Bird songs are used to attract mates and defend a territory. For most bird species, these are the roles of the male bird, so males do most of the singing. Bird songs are more common during the breeding season. The songs usually have a repeated pattern. They are typically longer and more complex than bird calls. Male and female birds both make calls year-round. The calls are used to keep in contact with other birds of the same species. They may be used to warn other birds of predators or to defend their territories. Baby birds make calls to get attention from their parents and demand food.

Northern mockingbirds can be tricky to identify by sound because they mimic the songs of other birds.

Western screech owls make a series of short hoots that get faster and faster. Their songs have been compared to the sound of a bouncing ball.

It takes time to learn the different songs and calls. Experts, such as park rangers, teach people about the birds in an area. People can listen to audio recordings of bird songs and calls to learn the sounds. Bird songs and calls vary in pitch, rhythm, tone, and repetition. Paying attention to these characteristics helps people identify songs and calls.

DID YOU KNOW?

Some birds are named for their calls. Chickadees, for instance, make a call that sounds like "chickadee-dee-dee."

FINDING AND IDENTIFYING ANIMALS

Woodpeckers make special sounds to mark their territories and attract mates. These birds drum on surfaces including trees, homes, and utility poles. Both males and females drum, and drumming occurs most often during the morning. Drumming can be heard from far away. The rhythm and drumming pattern varies by species, which allows birders another opportunity for bird identification.

With a wingspan of 30 inches (76 cm), the pileated woodpecker is the largest woodpecker in the United States.

BIG LISTERS

Some birders keep a list of the different bird species they have seen. They may log additional information, such as where and when they have first seen a species. Some birders have lists of dozens, hundreds, or even thousands of birds. People who have seen a long list of birds are referred to as Big Listers. These birders travel the world looking for different species in a variety of habitats. There are more than 10,000 bird species worldwide, and a few of the Big Listers have seen more than 7,000 of them.

IDENTIFYING AN ANIMAL

Location, habitat, season, and time of day are all factors that aid animal identification. Knowledge of tracks, scat, and sounds also provide clues about the type of animals nearby. After spotting wildlife, people should pay attention the size, color, and markings of the animal. These visual features can make it easier to identify an animal. People can also observe the animal's behavior, which can help with identification.

The shape and color of a moth's wing can help with identification. Luna moths are green and have wings with long tails.

FINDING AND IDENTIFYING ANIMALS

People may want to identify the fish they see while snorkeling. As with other forms of wildlife, the color, shape, and size of a fish help wildlife watchers with identification. Viewers should also pay attention to any patterns, such as stripes or spots, on the fish's body. Fish species have differently shaped mouths. Some have visible teeth. Wildlife watchers can also count the number of fins the fish has.

The white-banded triggerfish can be identified by its colorful bands and long snout.

WOLVES AND COYOTES

WOLF
30 inches (76.2 cm) tall at the shoulder
Weighs up to 100 pounds (45.4 kg)
Rounded ears
Rectangular face

COYOTE
18 inches (45.7 cm) tall at the shoulder
Weighs up to 45 pounds (20.4 kg)
Pointed ears
Long, narrow face

Wolves and coyotes overlap in habitat and look similar. But wildlife watchers can pay attention to size and the shape of the face to tell the difference between these two species.

Some animals are easily identifiable, while others are more challenging. This is especially true when animal species look similar to one another. For example, wolves and coyotes often share habitats and have similar appearances. One way to distinguish between them is by size. Coyotes are smaller. They weigh up to 45 pounds (20.4 kg) and stand 18 inches (45.7 cm) tall at the shoulder. Adult wolves weigh up to 100 pounds (45.4 kg) and stand 30 inches (76.2 cm) tall at the shoulder. Coyotes have snouts and ears that are narrower and pointier than those of wolves.

FINDING AND IDENTIFYING ANIMALS

Finding animals and making a confident identification takes time and practice. Guidebooks are useful tools to carry on a wildlife watching trip. They help people identify animals and look for signs of wildlife. People can also download wildlife apps to their smartphones. These apps provide a variety of information on wildlife, scat, tracks, and more.

Wildlife watchers can bring a guidebook with them to help with identification.

LEARNING MORE

Whether a wildlife watching trip is to a local park or out in the backcountry, there are people with expert knowledge of the area. Rangers, park volunteers, and other wildlife experts help inform decisions about where to go to see certain animals. They may point out places to avoid because of dangerous conditions.

Stopping by a visitor center is a great way to begin a wildlife watching trip. It can also be a destination at the end of a trip when people have questions about animal identification. Photos and descriptions of the animal and the location where it was spotted provide experts with the information they need for identification. In addition, many visitor centers have informational posters and displays for people to learn more about the animals in an area.

Visitor centers often have information about the wildlife in the area.

FINDING AND IDENTIFYING ANIMALS

Further, there are many groups with knowledge of specific animals. They provide information about wildlife and tips for viewing. Birders may want to consult with members of the National Audubon Society. This organization protects birds and their habitats. It also works to inform the public about birds and conservation. State or local chapters can provide information about birds in a specific area. Many regions have

Members of the Tulare County Audubon Society gathered to go birding in 2018. People who are interested in learning more about birdwatching can see if there is a local Audubon chapter near where they live.

a herpetological society that provides information about local reptiles. Herpetology is the study of reptiles. The North American Butterfly Association can assist butterfly enthusiasts. Some wildlife organizations offer guided trips, workshops, and lectures that teach wildlife watchers more about individual species and how to identify them. These organizations also allow people to meet others with similar interests.

WILDLIFE WATCHING SAFETY

Every wildlife watching trip comes with risk. Wild animals are unpredictable. Weather conditions can change, and injuries may occur to people spending time outdoors. To lower the chance of emergency situations, people should plan their trips ahead of time. Wildlife watchers should also bring other people with them on trips. This not only allows people to share wildlife watching experiences together but also helps keep them safe. If an emergency occurs, other people in the group can respond and get help.

Planning ahead for a wildlife watching trip reduces the risk of emergencies.

Before leaving home, wildlife watchers should make sure they have everything they need for a trip.

Before heading out on a trip, wildlife watchers should share trip information with a friend or family member. This includes information about the destination of the trip and how long it is expected to last. They should also share details about how they are arriving at the destination. For example, if wildlife watchers are driving, they should tell the friend or family member the make and model of the car and its license plate number. If people do not return from the trip at the expected time, the friend or family member will know to call for help. This information helps search and rescue teams find wildlife watchers who are missing.

WILDLIFE WATCHING SAFETY

Water trips may require additional planning for safety purposes. People need to make sure they have the proper gear. Some outings may be on still water, such as a lake or pond. Other excursions may be on fast-moving rivers. Wildlife watchers need to know how to handle rough waters. People planning to travel on the ocean need to consider currents and tides. All water trips require planning for launching and landing sites.

Always wear a life jacket during trips on water.

Trail markers help with navigation.

GETTING LOST

To avoid getting lost, everyone should stay on trails when possible. Maps should be checked frequently. Still, wildlife watchers may take a wrong turn and get disoriented. If they get lost, people should try not to panic. Panic results in poor decision-making and moving farther off course. The next step is to look at landmarks. A map and a compass help determine location. Being on a trail improves the likelihood that others will walk by and offer directional assistance. When there is cell coverage, individuals can call for help.

WILDLIFE WATCHING SAFETY

In some situations, the best plan is to sit in one place. Moving can result in getting farther off course. If it is almost night or if bad weather is approaching, the best course of action is to look for or build a shelter. While waiting, people need to stay fed, hydrated, and warm.

Giving three blasts on a whistle is a way to signal that there has been an emergency and someone needs help.

Large, dark clouds and a sudden drop in temperature are signs that a storm is approaching.

WEATHER CONDITIONS

People should pay attention to the weather forecast during a wildlife watching trip. The weather can change quickly in many wilderness areas and at high altitudes. Dry, sunny days can easily become cold, rainy afternoons. Storms can develop quickly. Mountains can block the view of the sky, making it difficult to spot an approaching storm. The onset of any storm requires rethinking a trip. Trips may need to be delayed or postponed. If people are already out, it may be necessary to return to a vehicle or to seek immediate shelter wherever possible.

WILDLIFE WATCHING SAFETY

In periods of drought or dry weather, lightning can cause wildfires.

Thunderstorms require extra caution. Lightning can be very dangerous. It can cause electrocution, resulting in injuries or death. Electrocution occurs when someone is struck by lightning. Electrical currents from lightning can also spread through the ground. People are electrocuted this way as well. If a thunderstorm is approaching, people should find shelter immediately.

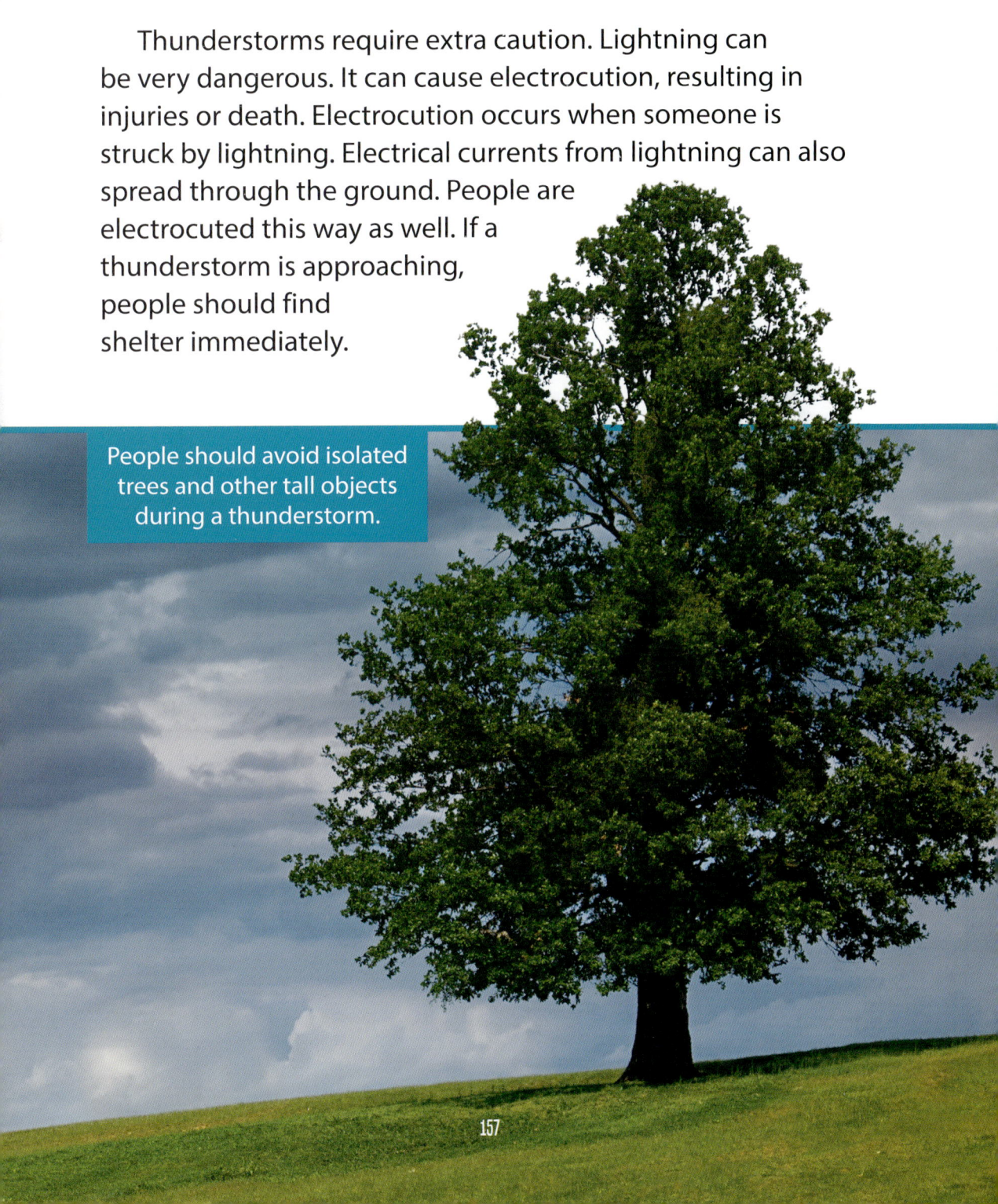

People should avoid isolated trees and other tall objects during a thunderstorm.

WILDLIFE WATCHING SAFETY

People are more likely to get electrocuted by lightning in or near open water than on land.

During a thunderstorm, people should drop all metal objects and move away from them. That includes umbrellas and backpacks with metal frames. Bicycles should be set aside. People on the ocean, a lake, or a river need to get off the water immediately. They should move at least 300 feet (91 m) away from the shoreline. They should also move to a lower elevation. Lightning tends to strike the tallest object in an area, so people should stay away from a single tall tree and avoid open areas where they are taller than their surroundings.

Shallow caves and overhangs are also places to avoid. The current from lightning can jump across the opening. This may jolt whoever is near that opening. The best place to take shelter during a thunderstorm is in a forested area at a low elevation where trees are roughly the same height. After taking shelter, everyone should spread apart and crouch down. They should stay low, but they should not lie down. Lying down increases contact with the ground. This increases the risk of electrocution from a ground current.

WILDLIFE WATCHING SAFETY

Wind is another element to be aware of. Strong winds can blow down trees and branches, causing injury or blocking travel routes. Strong winds can make travel difficult or impossible on oceans, lakes, or rivers. Before setting out, boaters should have a plan for alternative take-out sites in

> It is possible to tell if a thunderstorm is moving nearer by counting the seconds between a flash of lightning and thunder. If the length of time is decreasing, the storm is approaching.

case they cannot reach their destination. If possible, boaters should cut a trip short when winds increase. If they become stuck on the water, they should face boats into the wind. This helps boaters maintain the most control and avoid capsizing. Everyone should wear a life jacket.

ELEVATION

Elevation is an important consideration on a wildlife watching trip. There is less oxygen at higher elevations, which can decrease energy levels and make physical activity more challenging. At elevations above 8,000 feet (2,440 m), people may experience altitude sickness because of the low levels of oxygen. Common symptoms of altitude sickness include a headache, feeling dizzy, and shortness of breath. Symptoms range from mild to severe and can occur without any physical exertion.

To prevent altitude sickness, visitors to higher elevations should drink plenty of water, eat plenty of carbohydrates, and ascend slowly. People who are experiencing symptoms should stop physical activity. If symptoms do not subside, they should return to a lower elevation.

The summit of Pikes Peak in Colorado is reachable by tram and car. At 14,115 feet (4,302 m) in elevation, people are at high risk of experiencing altitude sickness.

COLD

Hypothermia may occur when wildlife watching. This happens when a person's body temperature drops below 95 degrees Fahrenheit (35°C). The body loses heat faster than it is able to produce it. This leads to symptoms such as shivering, shallow breathing, and confusion. Hypothermia can be deadly, so people should seek medical attention if they experience these symptoms.

Hypothermia usually occurs in cold weather, but people may experience symptoms even if temperatures are not below freezing. Wearing damp clothing can also cause the body to lose heat quickly. Clothing may become wet from rain or from sweat. People should wear clothing that dries quickly. They should

Warm winter hats can help prevent people from losing heat through their heads.

WILDLIFE WATCHING SAFETY

also pack extra layers when hiking to prevent hypothermia. Changing out of wet clothing into an extra change of clothes reduces the risk of hypothermia.

Frostbite is another health condition that occurs in cold weather. This is when the skin freezes. In severe cases of frostbite, deeper layers of the skin are also affected. As frostbite sets in, the skin becomes numb and may become discolored. Certain areas of the body are more vulnerable to frostbite. The hands and feet lose heat quickly, making frostbite more likely on these parts of the body. Exposed areas of skin, such as the ears, nose, and cheeks, are frequently affected.

Hypothermia may cause someone to feel tired and disoriented.

The earliest stage of frostbite is called frostnip. The skin becomes cold and numb.

People can reduce the risk of frostbite by wearing layers when wildlife watching in cold weather. A hat, gloves, and thick socks are essential. People should minimize the amount of exposed skin. The risk for frostbite increases as temperatures drop. At −18 degrees Fahrenheit (−28°C), people can develop frostbite in less than 30 minutes. They should move indoors when there is wind chill and extremely low temperatures.

Water-resistant gloves reduce the risk of frostbite. However, the insides of the gloves can become wet from sweat or melted snow, which causes a loss of body heat.

HEAT

When wildlife watching in hot weather, it is important to stay hydrated. People should make sure they have enough water with them for their trip. They should drink water at regular intervals. Hot weather and intense physical activity can cause people to become dehydrated. Wearing the correct clothing is also important when it is hot. Wildlife watchers should wear a hat. Cool, loose-fitting clothing keeps the sun off the skin. People can also rest in the shade to prevent overheating.

Heat-related illnesses include heat exhaustion. An individual who has a headache or feels nauseous should rest and drink water. He or she should try to cool down. In severe cases, people may experience heatstroke. This condition can be life-threatening, so people should seek medical attention as quickly as possible.

Symptoms of heat exhaustion include a headache, thirst, and feeling weak or nauseous.

UNDERSTANDING ANIMAL BEHAVIOR

Dangerous situations involving wildlife are rare. But wildlife watchers still must be cautious. Animals can behave unpredictably. Wildlife watchers need to have a basic understanding of animal behavior. They should be able to identify signs that an animal is stressed.

Keeping a safe distance from animals helps ensure that wildlife are not stressed by a person's presence.

WILDLIFE WATCHING SAFETY

Prairie dogs make different alarm calls depending on the type of predator they see.

Stress responses vary by species. Some responses may be vocal. Animals may make repeated chirps, squeaks, or grunts to indicate alarm. For example, prairie dogs bark as people or predators approach. This communicates to other prairie dogs that danger is near.

Wildlife watchers can also look for visual signs that animals feel threatened. Some animals raise their heads and look directly at viewers. Other animals may look visibly nervous or skittish. They may appear tense or make repeated vocalizations when stressed. Some animals pace or flap repeatedly. In rare circumstances, an animal may become aggressive. People should pay attention to the animal's body language. It may lower its head. The hairs on its shoulders and neck area may be raised. The animal may be preparing to charge toward the wildlife watcher. People must be prepared to respond in these situations.

Spotted skunks will often do a handstand as a warning sign before they spray.

SKUNKS

Skunks are known for the foul-smelling spray they emit when threatened. However, skunks often give warning signs before spraying. They may stamp their front feet and hiss. They may lunge forward short distances. Skunks may turn their backsides toward the threat and raise their tails. This is another warning sign. People who see these behaviors from skunks should stay calm and back away slowly.

WILDLIFE WATCHING SAFETY

WHAT TO DO IF APPROACHED, ATTACKED, OR BITTEN

Most animals want to keep their distance from humans. But animals may approach people because they are curious or looking for food. If people are approached by wildlife, they need to remain calm. They should back away slowly and not run. Wildlife watchers need to give the animal space so that it has room to escape. The animal likely wants to get away from people as much as people want to get away from it.

Feeding wildlife increases the risk of dangerous animal encounters.

People should not scream if they see a bear. High-pitched noises are similar to the sounds that prey make.

Black bears and grizzly bears are the most common bear species in the United States. Both kinds of bears pose great risk to humans. To prevent being approached or attacked by bears, people should give these animals plenty of space. If wildlife watchers notice that a female bear has cubs with her, they should be even more cautious. Females with cubs are very protective and may become aggressive. People should never get between a mother bear and her cubs.

WILDLIFE WATCHING SAFETY

Many wildlife watchers traveling in bear country carry bear spray with them. The spray is aimed at bears during an encounter. It should be easy to reach. Bear spray works at a distance of 60 feet (18 m). It contains chemicals that irritate the bear's eyes and airways. The spray is used only on a bear, not on clothing or skin.

Bear spray expires after about three to five years.

Black bears are smaller than grizzly bears.

If wildlife watchers encounter a bear, they should slowly back away from it. They should watch the bear but avoid eye contact. If the bear continues to approach, wildlife watchers need to be prepared to take additional steps to protect themselves. These steps differ depending on whether they are approached by a black bear or a grizzly bear. If a black bear approaches, people should try to make themselves appear as large as possible. They should hold their arms above their heads and wave sticks around. They should also make as much noise as possible to scare the bear away. If attacked, people should fight back. Punching the animal in the face and using nearby rocks or sticks as weapons may cause the bear to retreat.

WILDLIFE WATCHING SAFETY

With grizzly bears, wildlife watchers should not appear threatening. They should back away slowly and speak in a friendly tone. People should not run away from a grizzly bear or other predator. This can trigger a predator's instinct to attack. If the grizzly attacks, wildlife watchers should play dead. They should lie flat on their stomachs. They should use their hands to cover the backs of their necks.

Hikers can make noise to avoid surprising bears. A startled bear is more likely to attack.

Mountain lions can be very protective of their young.

For most predators, it is safest for people to stand their ground and appear threatening if approached. For example, if approached by a mountain lion, people need to make themselves look as big as possible by raising their arms and gear. They can also wave sticks or use other items to defend themselves. If the mountain lion does not back off, people need to show the lion that they are not prey. They should throw sticks or rocks at the animal. They should yell. If the lion attacks, people should continue to fight and try to stay on their feet. Sticks, rocks, water bottles, and other natural objects or gear can be used as weapons.

WILDLIFE WATCHING SAFETY

Alligators are predators that people should run away from. These animals are ambush predators. They stay still and strike when prey gets close. For this reason, it is safest for people

to run in a straight line as fast as possible if approached by an alligator. While alligators are quick, they cannot maintain top speed for very long. People are likely able to outrun an alligator.

Alligators can reach top speeds of 35 miles per hour (56 kmh) on dry land.

WILDLIFE WATCHING SAFETY

In the water, snorkelers and divers should stay alert for sharks. If approached, swimmers should stay calm. Sharks may confuse flailing and splashing for injured prey. Swimmers should maintain eye contact with the shark and slowly move away. If attacked, people should fight back. They should punch and kick the shark and poke its eyes or gills. Though sharks are not usually a threat to humans, people can take precautions to avoid them. One way is to avoid swimming at dawn and dusk when sharks are feeding. Swimmers should also avoid murky water and deep channels.

Tiger sharks are one of the few sharks known to attack people.

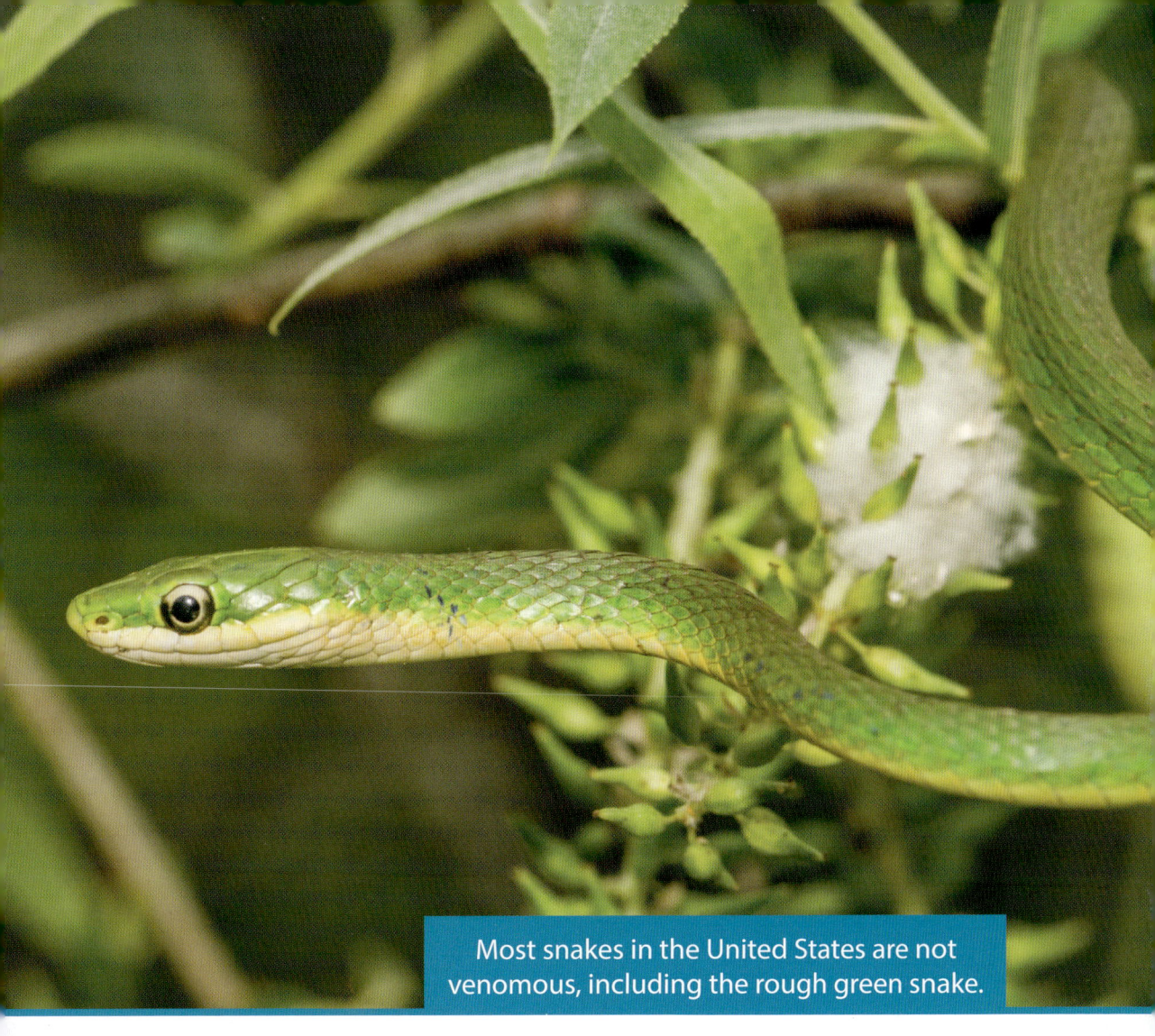

Most snakes in the United States are not venomous, including the rough green snake.

In general, snakes do not attack unless provoked. Most attacks occur because a person has disturbed a snake by accident. If people spot a snake, they should back away slowly. They need to make sure the snake has an escape route. Snakes may bite if they feel threatened. This is not only painful but also dangerous if the snake is venomous.

If bitten, people need to stay calm and remain still. If the snake is venomous, movement speeds up how quickly venom

WILDLIFE WATCHING SAFETY

MILK SNAKE OR CORAL SNAKE?

MILK SNAKE
Red touches black

CORAL SNAKE
Red touches yellow

Milk snakes are not venomous. But they look very similar to coral snakes, which are venomous. People can tell the difference between the two by looking at the color pattern of the bands.

is absorbed into the body. The area with the wound should be kept below the victim's heart. The wound needs to be cleaned and covered. Anyone bitten by a snake should seek immediate medical attention. If possible, people should take a photo of the snake or remember details about its appearance. This helps doctors determine if the snake was venomous.

Large prey animals, such as bison, are most likely to attack when they feel threatened. For this reason, people should not try to scare away bison or appear aggressive if one approaches.

Certain behaviors can indicate whether a bison feels threatened. The bison may snort and paw at the ground. It may also toss its head or raise its tail. If people see these actions, they should back away and give the bison plenty of space. They should prepare for the bison to charge. People can hide behind boulders, trees, or cars for protection.

Even small prey animals are unpredictable and may become aggressive. These animals may bite or scratch. Anyone who has contact with or is bitten by an animal should clean the site of the wound and seek medical attention. Animals may carry diseases such as rabies. Mammals including skunks, bats, foxes, and raccoons may have rabies. A rabid animal may behave strangely or aggressively. It may drool and appear anxious.

The big brown bat is one bat species that is known to transmit rabies to people.

But often, people cannot tell when an animal is sick. Rabies can be treated with a vaccine. Doctors determine if the medication is necessary. In humans, the first symptoms of rabies include fever, weakness, and headaches. Left untreated, rabies is often fatal.

Hantavirus is another virus wildlife watchers should be aware of. Rodents are the main carriers of this virus. It spreads to humans when people come in contact with an infected

Deer mice can be carriers of hantavirus.

animal's scat, urine, or saliva. It can also be transmitted through being bitten by an infected animal. Symptoms of hantavirus appear between one and eight weeks after exposure. Early signs include aches, fever, and fatigue. This may progress to coughing and difficulty breathing. People who have these symptoms should see a doctor.

Insects and arachnids can also be dangerous. Wearing long pants and long sleeves helps prevent stings or bites. People should wear bug repellant in areas with mosquitoes. Mosquito bites can be itchy, and some mosquitoes carry diseases. West Nile virus and Zika are two common mosquito-borne diseases in the United States. People who have been bitten by a mosquito and experience fever, headaches, joint pain, or other symptoms should seek medical attention.

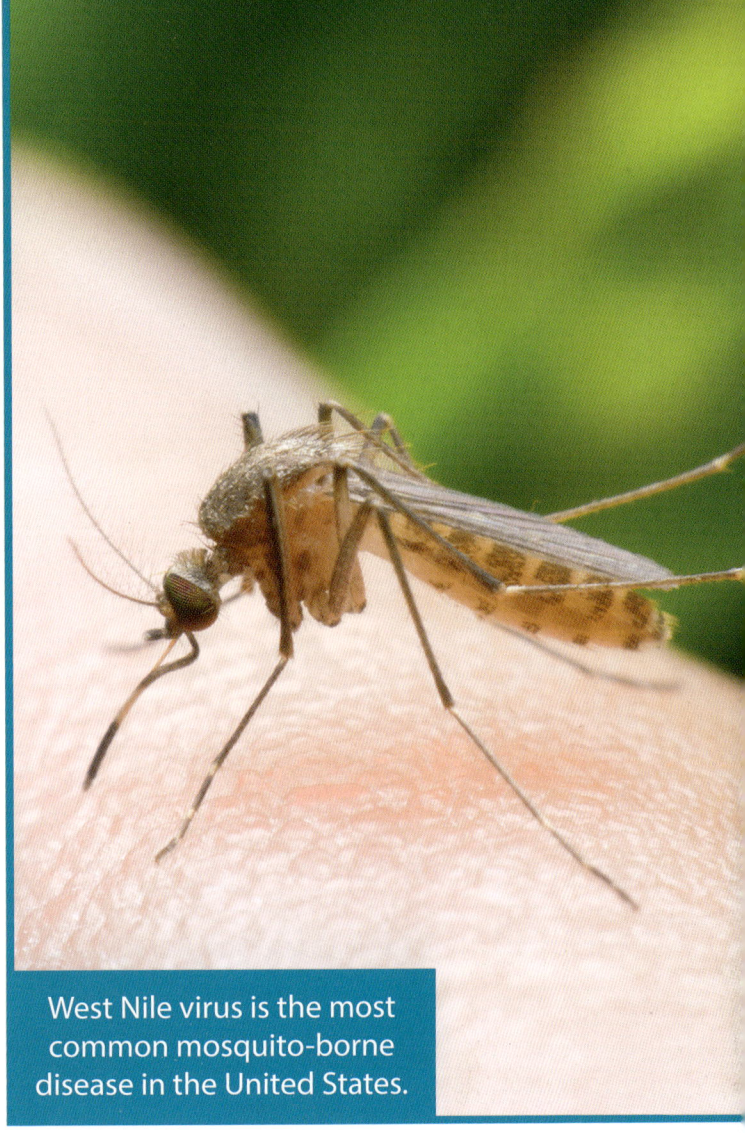

West Nile virus is the most common mosquito-borne disease in the United States.

WILDLIFE WATCHING SAFETY

Stings from bees and wasps are painful. Some people are allergic to these stings. A severe allergic reaction can be fatal. People who have a known allergy should carry emergency medications with them.

Under a magnification device, people can see a honeybee's barbed stinger. Because of this shape, the stinger becomes stuck in a person's skin.

A tick typically needs to feed on a host for 36 hours before it can transmit Lyme disease.

Ticks can spread diseases including Lyme disease and Rocky Mountain spotted fever. These diseases cause a rash, fever, chills, and other symptoms. The symptoms occur about three weeks after a tick bite. People should see a doctor if they think they have one of these diseases. Covering exposed skin and wearing bug repellent helps prevent tick bites. After a trip, people should check their bodies, clothes, and gear for ticks. If a tick is attached to skin, the tick needs to be carefully removed with tweezers. The area should also be cleaned.

WILDLIFE WATCHING SAFETY

While rare, spider bites can occur in urban and wilderness areas. Often people do not know they have been bitten. The first signs of a spider bite include redness and swelling of the skin. The area around the bite may be sore. However, some spiders are venomous, and reactions to their bites are more severe. Brown recluse and black widow spiders are two types of venomous spiders found in the United States. After being bitten by these spiders, people may experience migraines, muscle cramps, and even trouble breathing. Medical attention is necessary in these cases.

A black widow spider can be identified by a red hourglass marking on its abdomen.

Wildlife watching gives people the opportunity to connect with nature.

Dangerous interactions with wildlife are rare. But wildlife watchers still need to be ready for these encounters. Preparing for trips to the outdoors includes packing the proper gear. For a successful trip, wildlife watchers should have the resources and skill to find and identify animals. Wildlife watching is an activity for people of all ages and abilities. It is a chance to connect with nature. Many people enjoy seeing wild animals close to home and in the remote wilderness.

GLOSSARY

adapt
To become adjusted to a certain environment.

backcountry
A rural area or wilderness.

biodiversity
The number of different plant and animal species within an ecosystem.

conservation
The protection and management of natural resources.

contaminated
To be made dirty, infected, or unclean.

euthanize
To humanely put an animal to death when it is sick or because it is a risk to people.

forage
To search for food.

hibernation
A period of deep sleep or low activity to survive cold winter months.

insulation
Material that traps heat and is used for warmth.

migration
The process of traveling from one region to another to find food, mate, or give birth, typically seasonally.

pothole
A bowl-shaped depression eroded in a rock that may contain water.

refuge
A place or region that protects plants and wildlife.

spawning ground
The region or area where animals, such as fish, return to lay their eggs.

terrain
The physical features of an area of land.

topographic
Showing the physical features of an area including elevation, vegetation, and water.

TO LEARN MORE

FURTHER READINGS

Carson, Mary Kay. *Outdoor School: Animal Watching*. Odd Dot, 2021.

Perdew, Laura. *Mammals*. Abdo, 2021.

Vidal, Alexander. *Wilds of the United States: The Animals' Survival Field Guide*. Chronicle, 2022.

ONLINE RESOURCES

To learn more about wildlife watching, please visit **abdobooklinks.com** or scan this QR code. These links are routinely monitored and updated to provide the most current information available.

INDEX

altitude sickness, 162
apps, 13, 39–40, 146

bats, 66, 72, 85, 181
bear spray, 172
bears, 4, 22, 48, 58, 98, 116, 124–126, 133, 171–174
binoculars, 41–42
birds, 4–5, 7, 23, 33, 39–40, 42, 48, 53, 55, 77–78, 90, 92–93, 102, 106, 108, 110, 111, 118, 124, 131, 137, 140–142, 148
bison, 21, 51, 53, 81, 125–126, 180–181
burrows, 96–98, 130
butterflies, 5, 12, 102, 149

calls, 55, 104, 136–142
cameras, 47, 113
caribou, 81, 84, 110
citizen science, 11–13
clothing, 24, 30–34, 46, 66, 163–164, 166, 172, 185
coral reefs, 23, 62
crepuscular animals, 90, 103
cryptobiotic soils, 70

dehydration, 29, 166
diseases, 49–50, 54, 66, 73, 181–183, 185
diurnal animals, 86, 103
drinking water, 24, 28–29, 162, 166
durable surfaces, 67

edge habitats, 83
elk, 48, 83, 96, 104, 115, 133
erosion, 68–70, 77

field guides, 39–40
first aid kits, 43–44
fish, 4, 23, 72, 77, 85, 98, 110, 136, 144

fragile surfaces, 68–72
frogs, 72, 98, 136, 139
frostbite, 164–165

global positioning system (GPS), 38
guidebooks, 39, 146
guided tours, 7, 15, 17–19, 85

health benefits, 8
heatstroke, 166
hibernation, 66, 98–100
hot weather, 28, 102–103, 166
human food, 24–27, 56–59, 170
hypothermia, 30, 163–164

identification, 13, 15, 39–40, 42, 53, 112–113, 115–121, 125–128, 137, 139, 140–141, 143–147, 149, 187
iNaturalist, 40
insects, 4, 5, 7, 46, 78, 92–95, 100, 183
invasive species, 63–65

Leave No Trace, 53

maps, 24, 35–38, 153
mating, 52, 80, 100, 104–107, 136–137, 139–140, 142
mental health, 9
Merlin Bird ID, 40
migration, 12, 18, 80, 84, 97, 100, 109–110, 111
mountain lions, 48, 117, 119, 137, 175

National Audubon Society, 148
national parks, 7, 18–22, 35, 39–40, 48, 51, 81, 104
nocturnal animals, 88, 103

North American Butterfly Association, 149

park rangers, 18, 50, 52, 141, 147
plastics, 60
pollution, 59–62, 73

rain, 30, 32, 93–96, 122, 128, 155, 163

scat, 124–129, 143, 146, 183
sharks, 178
snakes, 99, 138, 179–180
snow, 32, 67, 96
spotting scopes, 41–42
stress to wildlife, 49–50, 54–55, 167–169
sunscreen, 45, 61–62

thunderstorms, 25, 93, 155–159
tracks, 115–124, 129, 143, 146

visitor centers, 35, 147

waste, 73–75
water trips, 18, 22–23, 31, 62, 65, 72, 75, 152, 158, 160–161, 178
white-nose syndrome, 66
wildlife corridors, 84–85
wildlife drives, 21–22
wind, 32, 91–92, 105, 114, 122, 160–161, 165

young animals, 52–53, 101, 107, 131, 137

zebra mussels, 65

PHOTO CREDITS

Cover Photos: Lars Poyansky/Shutterstock Images, front (tracks); Shutterstock Images, front (moose, field guide, cardinal, nest), back (binoculars); Ami Parikh/Shutterstock Images, front (blue jay); Miss Nuchwara Tongrit/Shutterstock Images, front (hiker); Tatiana Popova/Shutterstock Images, front (camera); Martin Bergsma/Shutterstock Images, front (trees); AR Pictures/Shutterstock Images, back (tourist binoculars)
Interior Photos: Roland Magnusson/Shutterstock Images, 1, 41 (top); Alex Cimbal/Shutterstock Images, 2–3, 20–21; Shutterstock Images, 4–5, 7, 14, 19, 25, 26–27, 30, 31, 32, 33, 35, 37, 38, 39, 41 (bottom), 42, 43, 44, 46, 47, 48, 49 (people), 49 (chipmunk), 49 (bighorn sheep), 49 (wolf), 54, 59, 64, 67, 73, 81, 86, 88–89, 90, 95, 99, 101, 102, 104, 105, 110–111, 113, 117, 118, 119, 120, 122–123, 127, 129, 130, 137, 139, 142, 144, 145 (left), 150, 151, 154, 157, 158–159, 165 (bottom), 164, 167, 174, 175, 178, 182, 183, 185, 186; Tory Kallman/Shutterstock Images, 5; Kristi Blokhin/Shutterstock Images, 6; Joanne Dale/Shutterstock Images, 8; Alexander Raths/Shutterstock Images, 9; Douglas Jennifer/Shutterstock Images, 10; John Morrison/E+/Getty Images, 10–11; Jason Whitman/NurPhoto/AP Images, 12; Gregory Rec/Portland Press Herald/Getty Images, 13; Dukas/Universal Images Group/Getty Images, 15; iStockphoto, 16, 40, 45, 55, 58, 112, 116, 136, 146, 166, 172; Mark Rightmire/Digital First Media/Orange County Register/MediaNews Group/Getty Images, 17; Matt Jeacock/iStockphoto, 18–19; Danita Delimont/Shutterstock Images, 22; M.M. Sweet/Moment/Getty Images, 23; Galyna Andrushko/Shutterstock Images, 24; Anatoliy Karlyuk/Shutterstock Images, 28; Przemek Klos/Shutterstock Images, 29; Tsveta Nesheva/Shutterstock Images, 34; Mike Flippo/Shutterstock Images, 36; Mikhail Romanov/Shutterstock Images, 50; Allan Vietmeier/Shutterstock Images, 51; Simon Evans/Shutterstock Images, 52; Jerry Larson/Waco Tribune-Herald/AP Images, 53; Cory Seamer/Shutterstock Images, 56; Nikolay Litov/Shutterstock Images, 57; Scuba Zoo/Science Source, 60; Martin Shields/Science Source, 61; A. L. Christensen/Moment Open/Getty Images, 62; Alamin Khan/Shutterstock Images, 63; Ed Reschke/Photodisc/Getty Images, 65; USFWS/Science Source, 66; Jordan Siemens/Stone/Getty Images, 68–69; Jon G. Fuller/VW Pics/AP Images, 70; Nature's Images/Science Source, 71; Edgar Lee Espe/Shutterstock Images, 72; Simon Vayro/Shutterstock Images, 74; Kelvin Wong/Shutterstock Images, 75; Jim Schwabel/Shutterstock Images, 76–77, 176–177; Karel Bock/Shutterstock Images, 78; Sophon Nawit/Shutterstock Images, 79; Joe Farah/Shutterstock Images, 80; Richard Seeley/Shutterstock Images, 82; Buvana Bala/Shutterstock Images, 83; Michael Warren/iStockphoto, 84–85; Jean-Edouard Rozey/Shutterstock Images, 87; Tom McHugh/Science Source, 88; Jiang Hongyan/Shutterstock Images, 91; Milan Zygmunt/Shutterstock Images, 92–93, 100; Valerie Giles/Science Source, 93; Jon Fuller/Visual & Written/SuperStock, 94; Mark Tierney/iStockphoto, 96; James P. Mock/Shutterstock Images, 96–97; Ian Stotesbury/500px/500 Px Plus/Getty Images, 98–99; Eric Isselee/Shutterstock Images, 103; Agnieszka Bacal/Shutterstock Images, 106, 169; Brian Lasenby/Shutterstock Images, 107, 134–135; Michael J. Cohen/Moment/Getty Images, 108; Steven Morello/SuperStock, 109; Evelyn D. Harrison/Shutterstock Images, 114; Chris Brannon/Shutterstock Images, 115; Kenneth Whitten/Irish Collection-Design Pics/SuperStock, 121; Amelia Martin/Shutterstock Images, 124; Lynn Bystrom/iStockphoto, 125; Steve Bower/Shutterstock Images, 126–127; Karel Stipek/Shutterstock Images, 128; Geoffrey Kuchera/Shutterstock Images, 131; John Morrison/iStockphoto, 132; Bryant Aardema-Bryants Wildlife Images/Moment/Getty Images, 133; Leena Robinson/Shutterstock Images, 135; Matt Jeppson/Shutterstock Images, 138–139; Raul Baena/Shutterstock Images, 140; Shravan Sundaram/Shutterstock Images, 141; Kevin Collison/Shutterstock Images, 143; Jim Cumming/Shutterstock Images, 145 (right); Stephanie Farrell/Shutterstock Images, 147; Chieko Hara/The Porterville Recorder/AP Images, 148–149; An Eduard/Shutterstock Images, 152; Yelizaveta Tomashevska/Shutterstock Images, 153; Tsikhanovich Alena/Shutterstock Images, 155; John D. Sirlin/Shutterstock Images, 156, 160–161; George Burba/Shutterstock Images, 162; Angelo Gilardelli/Shutterstock Images, 163; Chekalin Nikolai/Shutterstock Images, 165 (top); Thomas Stanton/Shutterstock Images, 168; Jeffrey Wickey/Shutterstock Images, 170; Chase Dekker/Shutterstock Images, 171; Menno Schaefer/Shutterstock Images, 173; Jay Ondreicka/Shutterstock Images, 179, 181; Petlin Dmitry/Shutterstock Images, 180 (left); Scott Delony/Shutterstock Images, 180 (right); Gary D. Gaugler/Science Source, 184; Dirk M. de Boer/Shutterstock Images, 187

ABDOBOOKS.COM

Published by Abdo Reference, a division of ABDO, PO Box 398166, Minneapolis, Minnesota 55439. Copyright © 2024 by Abdo Consulting Group, Inc. International copyrights reserved in all countries. No part of this book may be reproduced in any form without written permission from the publisher. Encyclopedias™ is a trademark and logo of Abdo Reference.

052023
092023

Editor: Angela Lim
Series Designer: Colleen McLaren
Production Designers: Cynthia Della-Rovere and Tara Raymo

LIBRARY OF CONGRESS CONTROL NUMBER: 2022949208

PUBLISHER'S CATALOGING-IN-PUBLICATION DATA

Names: Perdew, Laura, author.
Title: The wildlife watching encyclopedia / by Laura Perdew
Description: Minneapolis, Minnesota: Abdo Reference, 2024 | Series: Outdoor encyclopedias | Includes online resources and index.
Identifiers: ISBN 9781098291358 (lib. bdg.) | ISBN 9781098277536 (ebook)
Subjects: LCSH: Wildlife watching--Juvenile literature. | Tracking and trailing--Juvenile literature. | Animals--Juvenile literature. | Encyclopedias and dictionaries--Juvenile literature.
Classification: DDC 796.5--dc23